THE ESSENTIAL GUIDE TO
ALGEBRA 2

유하림(Harim Yoo) 지음

HERMONHOUSE

Preface

To. 학부모님과 학생들께

압구정 현장 강의를 통해 Prealgebra부터 AP Calculus까지 가르치면서, 강사로서 느낀 점은, 학부모님들이 선택할 수 있는 국내에서 판매되는 전문 교재가 몇 개 없다는 것과 충분히 많은 연구와 노력을 통해 개발된 교재들은 더더욱 부족하다는 점이었습니다.

국내 수능 시장에서는 "OOO" 커리큘럼으로 진행되는 수업들과 교재들이 많은 데 비해, 미국 SAT 시장과 유학 시장은 그렇지 않았기 때문에, 이렇게 유하림 커리큘럼의 시작을 알리는 것이 기대되면서도 떨립니다.

Algebra 2를 처음 배우는 학생들이 흥미롭게 배우길 희망하면서 교재를 썼습니다. 또한, 강의나 교재가 너무 쉽다고 느껴지진 않을 정도의 개념 강의 수준을 유지하기 위해 노력했으며, 지금까지 수업해왔던 내용들을 많이 담았기 때문에 국내에서 유학을 준비하는 학생이나 해외에서 공부하고 있는 학생들에게 도움이 되길 진심으로 희망해봅니다.

이 교재를 출간할 수 있도록 물심양면으로 힘써주신 마스터프랩 권주근 대표님께 감사합니다. 또한 저에게 항상 롤모델이 되어주고, 강사로서 성장할 수 있는 원동력을 주시고 계신 심현성 대표님께도 감사의 마음을 전합니다. 그리고 언제나 든든한 지원군인 제 아내와 딸, 부모님께도 항상 감사합니다. 마지막으로, 제 삶에 이러한 기회를 주신 하나님께 감사드립니다. 앞으로도 더 좋은 교재를 만들어 견고하고 튼튼한 유하림 커리큘럼을 완성시키길 희망합니다.

2020년 봄
유하림

저자 소개

유하림(Harim Yoo)

미국 Northwestern University,
B.A. in Mathematics and Economics
(노스웨스턴 대학교 수학과/경제학과 졸업)

마스터프렙 수학영역 대표강사
압구정 현장강의 ReachPrep 원장

고등학교 시절 문과였다가, 미국 노스웨스턴 대학교 학부 시절 재학 중 수학에 매료되어, Calculus 및 Multivariable Calculus 조교 활동 및 수학 강의 활동을 해온 문/이과를 아우르는 독특한 이력을 가진 강사이다. 현재 압구정 미국수학/과학전문학원으로 ReachPrep(리치프렙)을 운영 중이며, 미국 명문 보딩스쿨 학생들과 국내 외국인학교 및 국제학교 학생들을 꾸준히 지도하면서 명성을 쌓아가고 있다.

2010년 자기주도학습서인 "몰입공부"를 집필한 이후, 미국 중고교수학에 관심을 본격적으로 가지게 되었고, 현재 유하림커리큘럼 Essential Math Series를 집필하여, 압구정 현장강의 미국수학프리패스를 통해, 압도적으로 많은 학생들의 피드백을 통해, 발전적으로 교재 집필에 힘쓰고 있다.

마스터프렙 수학영역 대표강사 중 한 명으로 미국 수학 커리큘럼의 기초수학부터 경시수학까지 모두 영어와 한국어로 강의하면서, 실전 경험을 쌓아 그 전문성을 확고히 하고 있다.

[저 서] 몰입공부
The Essential Workbook for SAT Math Level 2
Essential Math Series 시리즈

저자직강 인터넷 강의 : 유학 분야 No.1 마스터프렙(www.masterprep.net)

이 책의 특징

유하림 커리큘럼 Essential Math Series의 핵심 교재 중 한 권입니다. Algebra 2를 배울 때 기본 개념과 그 응용 외에 무엇을 더 알아야만 학교 시험뿐 아니라 SAT 및 AMC 10/12 문제풀이까지 도달할 수 있을까 고민하면서 집필하였습니다. 특히 미국 명문 Boarding School 및 국내외 외국인학교에 다니는 9학년(10학년) 학생들이 반드시 숙지하고, 생각의 방향을 올바르게 키워나가서 결국 AMC와 같은 수학 경시대회에서도 우수한 성적을 거둘 수 있을까 고민을 많이 한 교재라고 단언컨대 말씀 드릴 수 있습니다

1st

본질에 가장 가까운 책

이 교재를 통해 학생들이 Algebra 2의 기본 내용과 그 응용뿐 아니라, 문제풀이의 본질에 하도록 유도하기 위해 집필하고 수정하는 과정을 반복했습니다. 부교재 역할이 아니라 핵심교재 역할을 하기 위해 만든 책이며, 현장강의와 인터넷 강의를 통한 학생들의 피드백을 받아 학생들의 눈높이에 맞춘 교재입니다.

2nd
생각의 확장을 위한 책

Essential Math Series를 AMC와 같은 경시수학을 준비하려는 학생을 위한 시리즈로 만들기 위해 예제 선정에 고민하고, 풀이방향을 잡았습니다. 특히, Algebra 2의 거의 많은 부분이 AMC 10/12와 같은 주요 수학경시대회에서 매우 주요한 역할을 한다는 점을 염두에 두고, 개념의 본질을 다양한 방법으로 확장시키기 위해, 여러 각도의 풀이가 가능한 문제들을 선별하였고, 심화 문제로 적용 가능한 풀이방법을 적용하는 문제 위주로 선정하였습니다. 학생들이 이 문제들을 하나하나 풀면서, 생각이 확장되는 것을 익히고, 더 깊은 생각을 할 수 있길 바랍니다

3rd
유학생을 위한 단 한 권의 책

Prealgebra부터 Algebra 2까지, 그리고 그 이후 수학 책도 마찬가지입니다. 미국수학을 정말 미국수학 답게 가르치기 위해 열심히 공부하고 연구하고, 앞으로도 그러할 것입니다. 노스웨스턴 대학교 학창시절 수학에 대한 열정을 뒤늦게 꽃피워 밤새워 공부했던 것처럼, 저는 학생들을 더 잘 가르치고, 더 나은 미래로 이끌기 위해, AMC, AIME, ARML, HMMT, PUMaC, SUMO와 같은 문제들을 동일한 열정으로 밤낮없이 풀고 해석합니다. 여러분이 지금 보는 이 책은 제 현재 노력의 최선의 산실이며, 앞으로도 그러할 것입니다. 이 책을 통해 수학을 두려워하지 않고, 문제 해결을 즐거워하며, 이른 나이에 수학에 대한 열정을 꽃피우길 기대합니다.

CONTENTS

저자직강 인터넷 강의는 SAT, AP No.1 인터넷 강의 사이트인 마스터프랩 (www.masterprep.net) 에서 보실 수 있습니다.

Topic 1

Real Number System

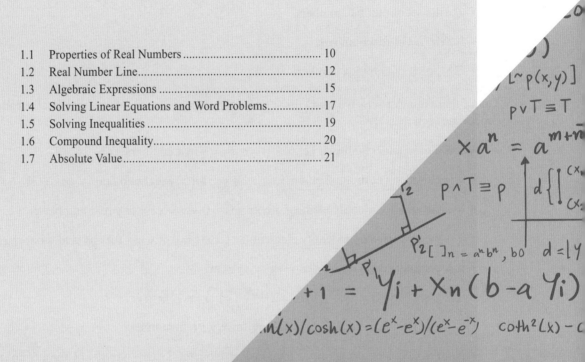

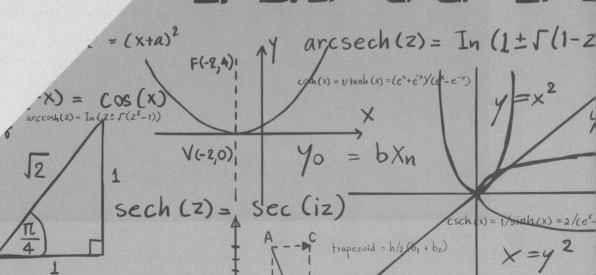

1.1　Properties of Real Numbers

$$\mathbb{N} \subset \mathbb{Z} \subset \mathbb{Q} \subset \mathbb{R} \subset \mathbb{C}$$

What are all these weird looking symbols? Well, they are the set of numbers we study in Algebra II.

- $\mathbb{N}$: the set of all natural numbers.

- $\mathbb{Z}$: the set of integers.

- $\mathbb{Q}$: the set of rational numbers.

- $\mathbb{R}$: the set of real numbers.

- $\mathbb{C}$: the set of complex numbers

Out of these number sets, we study the set of real numbers, i.e., $\mathbb{R}$, which satisfies the following properties such that

- it is <u>closed under addition</u>, i.e., $x + y = z$ for x, y real numbers $\implies z \in \mathbb{R}$.

- it is <u>closed under multiplication</u>, i.e., $x \times y = z$ for x, y real numbers $\implies z \in \mathbb{R}$.

- 0 is the <u>additive identity element</u>, i.e., $x + 0 = 0 + x = x$ for any real number x.

- 1 is the <u>multiplicative identity element</u>. i.e., $x \times 1 = 1 \times x = x$ for any real number x.

- it is <u>commutative</u>, i.e., $x + y = y + x$ or $xy = yx$.

- it is <u>associative</u>, i.e., $(x + y) + z = x + (y + z)$ or $(xy)z = x(yz)$

- it is <u>distributive</u>, i.e., $(x + y)z = xz + yz$ for x, y, z real numbers.

- Nonzero element has <u>multiplicative inverse element</u>, i.e., $x \times (1/x) = 1$ for any nonzero real x.

- Nonzero element has <u>additive inverse element</u>, i.e., $x + (-x) = 0$ for any real x.

Example

- If x and y are real numbers, $x + y = y + x$ because of commutative property.

- If x, y, and z are real numbers, $x(y + z) = xy + xz$ because of distributive property.

- If x, y, and z are real numbers, then $(x + y) + z = x + (y + z)$ because of associative property.

- The additive inverse of $-x$ is $-(-x) = x$.

- The multiplicative inverse of $1/x$ is $1/(1/x) = x$.

[1] What property of real number is demonstrated by the following equation?

$$3 + (2 + 4) = (3 + 2) + 4$$

(A) Identity Element of Addition
(B) Commutative Property of Addition
(C) Inverse Element of Addition
(D) Associatve Property of Addition

[2] What property of real number is demonstrated by the following equation?

$$3(2 + 4) = 3(2) + 3(4)$$

(A) Associative Property of Multiplication
(B) Commutative Property of Addition
(C) Distributive Property
(D) Closure under Multiplication

[3] Determine which set(s) of numbers, $\mathbb{N}, \mathbb{Z}, \mathbb{Q}, \mathbb{R}$ and $\mathbb{C}$, has an element of

(a) $-\sqrt{2}$ (b) -3

(c) $\dfrac{3}{4}$ (d) $\sqrt{-3}$

(e) $\sqrt[3]{-8}$

1.2 Real Number Line

The real number line is a straight line without any gap that has the distance notion. As one can see from the real number line in the figure below, we can *plot* the points, and compare its magnitude.

- 0 is called the <u>origin</u>.

- Numbers left of 0 are <u>negative</u>.

- Numbers right of 0 are <u>positive</u>.

The essential idea that we can learn from the real number line is that

- we can compare two numbers.

- we can CASEWORK when we deal with two real numbers.

In order to compare two real numbers, we first introduce the notion of magnitude associated to a real number x, i.e.,

$$|x - 0| = |x|$$

which is always greater than or equal to 0.

4 Arrange the following numbers in increasing order.

$\sqrt{2}$ $\qquad$ 2 $\qquad$ $\sqrt{5}$ $\qquad$ $\sqrt{3}$ $\qquad$ 1

Let's practice more with the radical expression. Given a real number line, it is important to find out which N value that satisfies $N \leq \sqrt{x} < N+1$. In other words, we will look at problem-solving strategy for finding out $N^2 \leq x < (N+1)^2$.

Moreover, we can casework depending on the given number. For instance, given a number L, we can find the distance between x and L depending on whether $x < L$ or $x > L$.

- if $x < L$, then $|x - L| = L - x$.

- if $x > L$, then $|x - L| = x - L$.

$\boxed{5}$

(a) Which letter on the graph corresponds to $\sqrt{3}$?

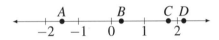

(A) A (B) B (C) C (D) D

(b) Which letter on the graph is closest to -0.5?

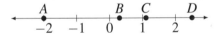

(A) A (B) B (C) C (D) D

6 What is the midpoint between -3 and 7?

(A) -4
(B) 4
(C) 2
(D) -2

7 Which number is farthest from -2 on a number line?

(A) 2 (B) $\dfrac{4}{3}$ (C) $-\dfrac{1}{2}$ (D) 3

8 Mark the points on the real number line with distance 4 from -1.

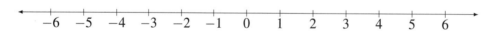

1.3 Algebraic Expressions

Given a word phrase, you could assign a variable and a constant. Think about which unknown we want to set as a variable. Often times, the question directly asks that unknown. Otherwise, we need to specifically choose the unknown to make the problem easier to set up and solve. As soon as we write an algebraic expression, we could perform two operations.

- We evaluate the algebraic expression by substitution.

- We simplify the algebraic expression by combining like terms.

Example

Part1. Evaluate $3x + 2y$ if $x = 1$ and $y = 2$.

Solution
Substituting $x = 1$ and $y = 2$ into the given algebraic expression, we get $3(1) + 2(2) = 3 + 4 = 7$.

Part2. Simplify the following algebraic expressions by combining the like terms.

$$3x^2 + 2y - x^2 - 4y$$

Solution
Since $3x^2$ and $-x^2$ are like terms, they are combined together into $2x^2$. Likewise, $2y$ and $-4y$ are like terms, so they are added to result in $-2y$. Hence, $2x^2 - 2y$ is the simplified form of the above algebraic expression.

9 Write an algebraic expression that models the following word problem : the sum of three consecutive integers is 72. Then, find the three numbers.

10 Write an algebraic expression that models the following word problem : A number is 4 less than 3 times the other number. If they sum to 44, find the two numbers.

11 A retail store has monthly fixed costs of $3,000 and monthly salary costs of $2,500 for each employee. If the store hires x employees for an entire year, write the equation that represents the total cost c, in dollars, for the year.

12 Write the following equation that represents the square of the sum of x and y, decreased by the product of x and y.

1.4 Solving Linear Equations and Word Problems

We usually solve linear equations by moving all constants to one side of the equation and the variable to the other side by subtraction and addition. Then we divide both sides by the coefficient of the variable.

Example

Solve $3 - 2(x - 3) = 11$.

Solution

$$3 - 2(x - 3) = 13$$
$$3 - 2x + 8 = 13$$
$$11 - 2x = 13$$
$$-2x = 2$$
$$x = -1$$

---

13 John is 7 years older than Jane. In 5 years, John will be twice as old as Jane. How old is John now?

14 William walks 60 meters per minute. David can walk 120 meters per minute for the first 10 minutes, but then slows down to 20 meters per minute thereafter. If they start walking at the same time, after how many minutes t will both William and David have walked the same distance?

15 If 5 is added to the square root of x, the result is 9. What is the value of $x+2$?

16 A grocery store sells tomatoes in boxes of 4 or 10. If Jenny buys x boxes of 4 and y boxes of 10, where $x \geq 1$ and $y \geq 1$, for a total of 60 tomatoes, what is one possible value of x?

17 If 3 is subtracted from 3 times the number x, the result is 21. What is the result when 8 is added to half of x?

1.5 Solving Inequalities

There are multiple types of inequalities we would learn in Algebra 2. The golden rule to remember when we solve the inequality is that we <u>reverse</u> the sign every time we multiply or divide by a negative number.

Example

Solve $-3x - 7 \leq -7x - 27$.

> **Solution**
>
> $$-3x - 7 \leq -7x - 27$$
> $$4x \leq -20$$
> $$x \leq -5.$$

---

18 Graph the solution of $x + 2 < 2x - 1$.

19 Tina must limit her daily sugar consumption to at most 40 grams. One cookie has 5 grams of sugar and one fruit salad contains 8 grams of sugar. If Tina ate only cookies and fruit salads, write the inequality that represents the possible number of cookies c and fruit salads s that she could eat in one day and remain within the limit.

1.6 Compound Inequality

There are two types of inequality : conjunctive inequality and disjunctive inequality.

- Conjunctive inequality indicates <u>and</u>.

- Disjunctive inequality indicates <u>or</u>

Example

Solve $x + 1 < 2$ or $2x - 1 > 3$.

Solution

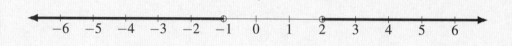

20 Solve $-5 < 3x - 2 < 11$.

21 Solve $2x + 3 < -7$ <u>or</u> $-2x - 1 < 7$.

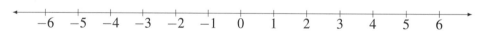

1.7   Absolute Value

The absolute value of x, i.e., $|x|$, is the distance x is from 0. In other words,

- If x is positive, $|x|$ stays positive.

- If x is negative, $|x|$ becomes positive.

Similarly, given two real numbers a and b, $|a-b|$ is the distance between a and b, so it is obvious that $|a-b| = |b-a|$.

Example

How many integer values of x satisfy $|x| < 4$?

Solution

$$-4 < x < 4$$
$$-3 \leq x \leq 3$$

Hence, there are 7 integer values.

22 If $x < y$, what is $|x-y| - (x-y)$?

(A) 0   (B) $2x - 2y$ (C) $2y - 2x$ (D) $x - y$

23 Solve $|x+1| = 3$. Plot the solutions and -1, and make observation out of the three points.

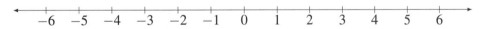

24 For which of the following values of x is

(a) $|2x - 5| < 0$?

(b) $|2x - 5| = 0$?

(c) $|2x - 5| > 0$?

25 If $|x - 2| = 10$, find the sum of the two possible values of x.

26 If $|x-1| = 2x+3$, solve for x.

(a) If $x \geq 1$, then $|x-1| = x-1 = 2x+3$. Solve for such x.

(b) If $x < 1$, then $|x-1| = 1-x = 2x+3$. Solve for such x.

27 If $|3x-4| = -2x+1$. Solve for x.

As we saw in the previous examples, if $|x - a| < b$, then a is the midpoint of the three numbers $a - b$, a, and $a + b$. We will use this idea in reverse direction to write down an absolute-valued inequality to express the given conjunctive (or disjunctive) inequality.

If $-2 < x < 6$, set up the absolute-valued inequality for x.

Solution
The midpoint of -2 and 6 is 2. So, $-2(= 2 - 4), 2, 6(= 2 + 4)$ are plotted in the real number line.

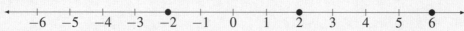

Hence, x is at most 4 away from -2, i.e., $|x - 2| < 4$.

28 A hot dog factory must ensure that its hot dogs are between $6\frac{1}{4}$ inches and $6\frac{3}{4}$ inches in length, including the endpoints. If h is the length of a hot dog from this factory, then find the absolute-valued inequality that correctly expresses the accepted values of h.

29 Rolls of tape must be made to a certain length. They must contain enough tape to cover between 400 feet and 410 feet, including the endpoints. If l is the length of a roll of tape that meets this requirement, find the absolute-valued inequality that expresses the possible values of l.

1 If a is the additive inverse of -2, and b is the multiplicative inverse of 0.75, then find ab.

2 Solve for x.

$$\frac{2-x}{x+1} + \frac{2x-4}{2-x} = 1$$

3 What is the value of the expression $\dfrac{a^2 - a - 6}{a - 3}$ for $a = 2$?

4 Evaluate $\dfrac{3 + b(3 + b) - 3^2}{b - 3 + b^2}$ for $b = -2$.

5 Which of the real number property explains $3x + xy = (3 + y)x$?

(A) Associative Property
(B) Commutative Property
(C) Identity Property
(D) Distributive Property

6 In April 2007, the price of a train started from $120 and increased by $20 each week.

(a) If t is the number of weeks since the beginning, write down the expression for the price of the train in terms of t.

(b) If Bob wanted to pay at most 170 dollars for the ticket, how long can he wait since April 2007?

$\boxed{7}$ What is the largest integer x satisfying $\dfrac{x}{3} + \dfrac{4}{5} < \dfrac{5}{3}$?

$\boxed{8}$ How many integers n satisfy both of the inequalities $4n + 3 < 25$ and $-7n + 5 < 24$ at the same time?

9 Solve the absolute-valued equation $|2-x| = 3+2x$.

10 If $3 < 2x-1 < 5$, graph the solution on the real number line and set up the absolute-valued inequality for x.

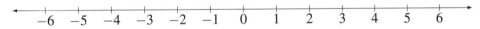

1 $ab = \dfrac{8}{3}$

2 $x = -\dfrac{1}{4}$

3 4

4 8

5 (D)

6 (a) $p(t) = 120 + 20t$ (b) 2.5 weeks

7 2, since $x < 2.6$.

8 There are 8 integers, i.e., $-2, -1, 0, 1, 2, 3, 4, 5$.

9 $x = -\dfrac{1}{3}$.

10 $\left| x - \dfrac{5}{2} \right| < \dfrac{1}{2}$ and the figure is drawn below.

Topic 2

Function and Linear Graph

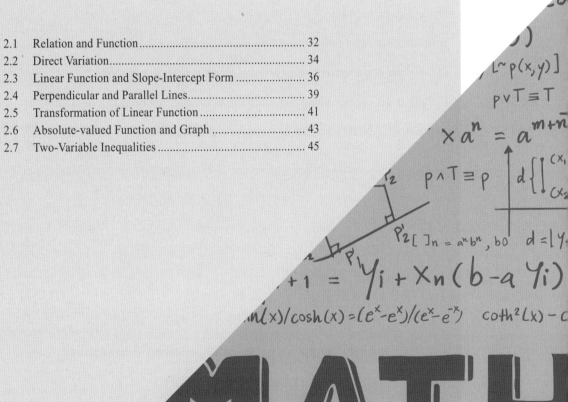

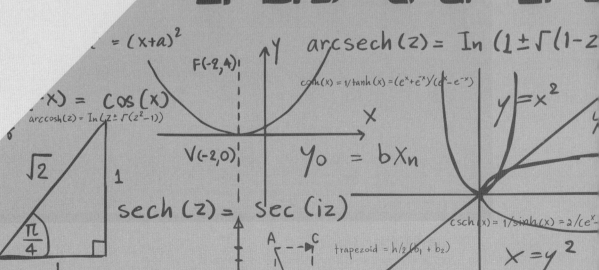

2.1 Relation and Function

Relation is a set of (x, y) pairs, where x and y are normally constants. For instance, $\{(1,2), (1,3), (3,3), (3,4)\}$ is a relation of x and y. On the other hand, function is a relation where each element in the domain is paired with exactly one element in range. In other words, a function relates domain (the set of x values) and range (the set of y values). For example, $\{(1,2), (2,3), (3,4), (4,3)\}$ is a function. For one value of x, there is at most 1 value of y.

How do we distinguish function from relation? The answer to this question can be easily answered by a test called "Vertical Line Test." This test, sometimes abbreviated as VLT, is a test to determine whether the graph belongs to a function or not. The procedure goes as

1. Draw any vertical line.

2. If there are at least two intersection points, the graph is not of a function.

3. If there are one intersection point, the graph is that of a function.

4. If there is no intersection point, then the associated value is not defined.

1 Which point CANNOT be part of a function that includes $(3, -1)$, $(4, 2)$, $(5, 4)$, $(-2, 0)$ and $(8, -3)$?

(A) $(6, -7)$ (B) $(3, -2)$ (C) $(7, 4)$ (D) $(2, 2)$

2 What is the domain of the relation given by the following ordered pairs?

$$(1,2), (2,1), (-2,-1), (3,1), (4,1)$$

(A) $\{-1, 1, 2\}$
(B) $\{1, 2\}$
(C) $\{-2, 1, 2, 3, 4\}$
(D) $\{-2, -1, 1, 2, 3, 4\}$

Usually, a function takes x as an input, but this is not always the case. Sometimes, it could take any other expression, even a function itself, as an input. Let's have a look at this phenomenon in the following example.

3 If $f(x) = 3x^2 + 4$, compute $f(2x+1)$.

In the following example, let's have a look at what happens when a function has another function as an input.

4 If $f(x) = 2x + 3$ and $g(x) = -x + 1$, find the values of m such that $f(g(m)) = 1$. (Think about how we know there exists any value of m satisfying $f(g(m)) = 1$.[1])

5 If $f(x) = x^x$, then what is the value of $f(1) + f(2) + f(3) + f(4)$?

---

[1] A linear function is known as a surjective function, meaning that if a bystander gives us a random output, we can always find an input associated to it.

2.2 Direct Variation

If x varies directly with y, then we say $x = ky$ where k is some constant. We call k as the constant of variation. The graph of (x, y) satisfying a direct variation must pass through the origin. In other words,

- $y = 3x + 1$ is NOT a direct variation.

- $y = -4x$ is a direct variation.

Example

Assume y varies directly as x, and if $y = 3$ when $x = -9$, find x when $y = 5$.

Solution

Let $y = kx$ for some constant k. Then, $3 = -9(k)$, so $k = -\frac{1}{3}$. Hence, $5 = -\frac{1}{3}(x)$, so $x = -15$.

---

6 The distance traveled by a cyclist varies directly with the number of pedal strokes. Suppose that in the bicycle's lowest gear, 6 pedal strokes move the cyclist about 240 inches. In the same gear, how many pedal strokes are needed for the cyclist to move 100 ft? (Hint : 1 ft = 12 inches)

7 The distance a spring stretches varies directly with the amount of weight that is hanging on it. A weight of 2.5 pounds stretches a spring 18 inches. What is the stretch of the spring when a weight of 6.4 pounds is hanging on it?

8 A new hybrid car has a full 12-gallon gas tank. On one tank of gas, the owner can drive 540 miles. The number of miles traveled by the car varies directly with the number of gallons of gas it burns.

(a) Write an equation that relates the number of miles traveled with the number of gallons of gas used.

(b) How many miles can the car owner travel using 9 gallons of gas?

9 On a certain calling plan, a 15-minute long-distance phone call costs $0.90. The cost varies directly with the length of the call. Write an equation that relates the cost to the length of the call. How long is a call that costs $1.32?

2.3   Linear Function and Slope-Intercept Form

Given a linear equation $y = mx + b$, m represents the slope and b represents the y-intercept. It is more important to know the meaning of the slope and the intercept in terms of question contexts.

- The slope is the unit change of y per unit change of x.

- The y-intercept is the y-value when $x = 0$.

10

$$h(t) = 200 - 4t$$

The equation above can be used to model the number of hours h until a gallon of gas at a temperature of t, in degrees Celsius, goes vaporized. Based on the model, which of the following is the best intrepretation of the number 4 in the equation?

(A) An increase of $1°$ C will make a gallon of gas go vaporized 4 hours faster.
(B) An increase of $1°$ C will make 4 gallons of gas go vaporized 1 hour faster.
(C) An increase of $4°$ C will make a gallon of gas go vaporized 1 hour faster.
(D) An increase of $4°$ C will make a gallon of gas go vaporized 4 hours faster.

11
A luxury car was sold at an auction. The price p of the car, in hundred dollars, during the auction can be modeled by the equation $p(t) = 50,000 - 100t$, where t is the number of seconds left in the auction. According to the model, what is the meaning of 50,000 in the equation?

(A) The initial auction price of the car
(B) The final auction price of the car
(C) The increase in the price of the car per second
(D) The time it took to auction off the car, in seconds

Recall that the linear equation is easily determined by two distinct points. In Geometry, we learned that two points *uniquely* determine the line graph that passes through the two points. Let's find out other points on the graph by utilizing coordinate geometry.

12 In the xy-plane, two points $(-3,5)$ and $(6,8)$ lie on a line. Which of the following points is also on the line?

(A) $(3,7)$ (B) $(0,4)$ (C) $(9,10)$ (D) $(12,15)$

13 What is true about the line that passes through the points $(3,-7)$ and $(3,2)$?

(A) It is horizontal.
(B) It is vertical.
(C) It rises from left to right.
(D) It falls from left to right.

14

$$\begin{cases} y = mx + b \\ y = -bx \end{cases}$$

The equations of two lines in the xy-plane are shown above, where m and b are real constants. If the two lines intersect at $(2,8)$, what is the value of m?

(A) 2   (B) 4   (C) 6   (D) 8

A linear function $y = f(x) = mx + b$ always has the graph of a line. This line cannot have a cusp, an abrupt turn, in its graph. In other words, if the graph of the function tends to go upward, then it must maintain its direction. Similarly, if the graph of the function tends to go downward, it must go downward, maintaining its slope. One feature we remember learning in Algebra 1 is that the linear function has a constant slope. The constant slope means that the rate of change of y per change of x NEVER changes, for two pairs of (x_1, y_1) and (x_2, y_2) satisfying the linear function.

Given $y = f(x) = mx + b$, regardless of the value of b, the graph's direction is determined by the sign of m.

- $m > 0$: the graph goes upward.

- $m = 0$: the graph is horizontal.

- $m < 0$: the graph goes downward.

15 If $f(x)$ is a linear function such that $f(0) \leq f(1)$ and $f(2) \geq f(3)$, and $f(10) = 4$, which of the following must be true?

(A) $f(4) < f(6) < f(5)$
(B) $f(1) = 0$
(C) $f(0) < 0$
(D) $f(0) = 4$

2.4 Perpendicular and Parallel Lines

Given a line equation $y = f(x) = mx + b$, there are possibly two important lines related to the given equation.

- Perpendicular line has a slope of $-\dfrac{1}{m}$, meaning that the slope is the negative reciprocal of the original one.

- Parallel line has a slope of m, meaning that the slope is unchanged.

$\boxed{16}$ Which of the following numbers correctly represents the slope of a line that is *parallel* to a line with a slope of -2?

(A) $-\dfrac{1}{2}$ (B) $\dfrac{1}{2}$ (C) -2 (D) 2

In Algebra 1, we focused on the form of $y = mx + b$, known as the slope-intercept form. On the other hand, if the line has the slope of m and passes through (x_1, y_1), then the equation must be in the form of

$$y = m(x - x_1) + y_1$$

$\boxed{17}$ Which of the following linear equations represents a line through $(-1, 1)$ with a slope of $2/3$?

(A) $y = \dfrac{2}{3}(x + 1) + 1$

(B) $y = \dfrac{2}{3}(x - 1) + 1$

(C) $y = \dfrac{2}{3}(x - 1) - 1$

(D) $y = \dfrac{2}{3}(x + 1) - 1$

18 The slope of line l is $\dfrac{1}{2}$ and its y-intercept is 3. What is the equation of the line perpendicular to the line l and passes through $(1,5)$?

19

$$y = \frac{a}{b}x + c$$
$$y = \frac{d}{e}x + f$$

The equations of two perpendicular lines in the xy-plane are shown above in the system of equations, where a, b, c, d, e and f are valid constants. If $0 < \dfrac{a}{b} < 1$, which of the following must be true?

(A) $\dfrac{d}{e} < -1$

(B) $-1 < \dfrac{d}{e} < 0$

(C) $0 < \dfrac{d}{e} < 1$

(D) $1 < \dfrac{d}{e}$

2.5 Transformation of Linear Function

There are four types of transformation. In this section, we will cover three transformations.

- Dilation : either stretching or shrinking.

  [1] $f(kx)$ changes the width, not height.

  [2] $kf(x)$ changes the height, not width.

- Reflection : Reflection about the x-axis, y-axis, and the origin.

  [1] $f(-x)$ reflects $f(x)$ about the y-axis.

  [2] $-f(x)$ reflects $f(x)$ about the x-axis.

  [3] $-f(-x)$ reflects $f(x)$ about the origin.

- Translation : Vertical shift or horizontal shift.

  [1] $f(x-h)$ shifts it right or left.

  [2] $f(x)+k$ shifts it up or down.

20 If $(6,4)$ is on the graph of $y = f(x)$, which of the following must be on the graph of $y = f(2x-2)+2$?

(A) $(6,4)$　　　　(B) $(4,6)$　　　　(C) $(3,4)$　　　　(D) $(5,6)$

21 If $(6,4)$ is on the graph of $y = f(x)$, which of the following must be on the graph of $y = f(2x)/2+2$?

(A) $(6,4)$　　　　(B) $(6,2)$　　　　(C) $(3,2)$　　　　(D) $(3,4)$

As shown in the previous two examples, when a function is transformed, it is best for us to consider dilation or reflection first. In the following example, we would like to see how $f(-x)$ plays a role in transformation.

Example

Explain the transformation of $y = f(x)$ into $y = f(1-x) + 2$.

Solution
First off, switch $f(1-x)$ into $f(-(x-1))$. Then, we notice that the graph of $y = f(x)$ is reflected about the y-axis. Perform translation as the final step. In other words, the reflected graph is shifted 1 unit to the right direction and 2 units to the upward direction.

---

22 If $(6,4)$ is on the graph of $y = f(x)$, which of the following must be on the graph of $y = f(2-x) + 2$?

(A) $(-6,4)$
(B) $(-6,2)$
(C) $(-4,4)$
(D) $(-4,6)$

23 The graph of $y = f(x)$ is reflected about the x-axis and translated 3 units right. Which of the following correctly describes the resulting equation?

(A) $y = -f(x+3)$
(B) $y = f(-x+3)$
(C) $y = -f(x-3)$
(D) $y = f(-x-3)$

2.6 Absolute-valued Function and Graph

The first piece-wise function we study is the absolute value function $y = |x|$ whose graph can be illustrated by the following diagram.

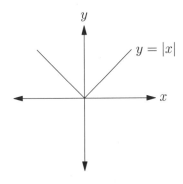

- The vertex is located at $(0,0)$. Normally, in the graph of $y = a|x - h| + k$, the vertex is located at (h,k).

- Due to the definition of $|x|$, we graph $y = |x|$ in two different cases : $x \geq 0$ or $x < 0$. When $x < 0$, the graph of $y = |x|$ is equivalent to $y = -x$. When $x \geq 0$, the graph of $y = |x|$ is equivalent to $y = x$.

24 Graph $y = -|x - 2| + 3$. Label the vertex, the x-intercepts, and y-intercept, if any.

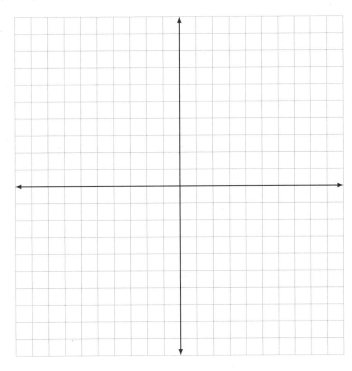

25 Graph the function $y = -\dfrac{1}{3}|x+2| - 2$ in the following xy-plane.

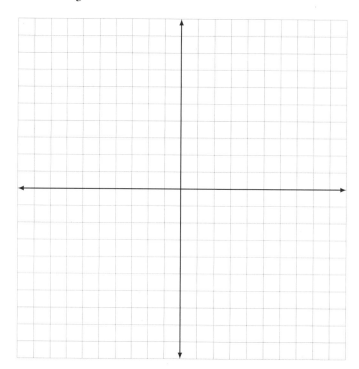

As one can see from the previous example, the graph of $y = a|x|$ is either a steep line or a wide line depending on the value of a. If $1 < a$, then the graph goes *steep*. If $0 < a < 1$, the graph goes *wide*.

26 The graph of which equation is the graph of $f(x) = |x|$ reflected about the x-axis, translated 2 units left, vertically compressed[2] by a factor of $\dfrac{1}{3}$, and translated up 4 units?

(A) $y = 3|x-2| + 4$

(B) $y = -3|x+2| + 4$

(C) $y = -\dfrac{1}{3}|x+2| + 4$

(D) $y = -\dfrac{1}{3}|x-2| + 4$

---

[2]Here, I use *vertical compression* as a literal sense of vertical shrinking. In this book, a vertical compression by a factor of $1/3$ means that the original height is truncated into its third.

2.7   Two-Variable Inequalities

In order to draw a linear inequality in two variables, apply the following procedures.

1. Find some table values satisfying the given inequality.

2. Draw a demarcation line (boundary line) by setting the inequality as equation. Here, the boundary line is either solid or dashed. If the inequality is inclusive, the line is solid. Otherwise, the line is dashed.

3. Find the plane in which the table values sit.

4. Color the boundary line (or dotted line) and the plane of interest.

Example

How many solution pairs are there for $y \leq x$?

Solution

A solution pair (x,y) for $y \leq x$ is the set of all possible (x,y)'s such that $y \leq x$. Given any two real numbers x, y, it is always true that either $x < y$, $x = y$ or $x > y$. Since we are looking for the two possible cases of the three possibilities, and there are infinitely many real numbers in $\mathbb{R}$, we conclude that there are infinitely many solution pairs satisfying $y \leq x$. In particular, there are solution pairs (x,y) such that $x = y$ or $y < x$.

27 Find the graph of $y \leq 3x$.

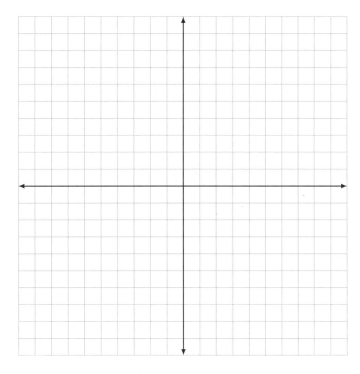

We should be a bit more careful when we solve word problems. Sometimes, as we set the variables, it is quite common for us to make mistakes about neglecting necessary constraints on the variables. For example, if we let x as the number of students, then $x \leq 0$ and x must be a whole number. Let's have a look at the following examples.

28 A high school band is expecting to take at least 120 students to a regional band competition. The school rents some passenger vans that can transport 8 students. Other students, in groups of 4, will need to ride in personal vehicles driven by parents. Write an inequality that shows every possible combination of vans and cars that could be used to drive students to the competition.

29 A salesperson sells two models of vacuum cleaners. One brand sells for $150 each and the other sells for $200 each. The salesperson has a weekly sales goal of selling at least $1800. Write an inequality relating the revenue from the vacuum cleaners to the sales goal.

1 If $f(x) = 3x - a$ and $f(1) = 2$, then find $f(3) - f(-2)$.

(a) Evaluate it directly.   (b) Evaluate it using the slope form.

2 If $(ab, a - b)$ is in the 3rd quadrant, find the quadrant which has the point symmetric to (a, b) with respect to the origin.

3 If the graph of a linear function $y = ax$ passes through $(6, 1)$ and $(12, b)$, then find the value of b.

4 Find the product of the slope and y-intercept of the equation $2x + 3y = 12$.

5 Sketch the graph of $4x + 2y = 16$, and label the two intercepts-x-intercept and y-intercept.

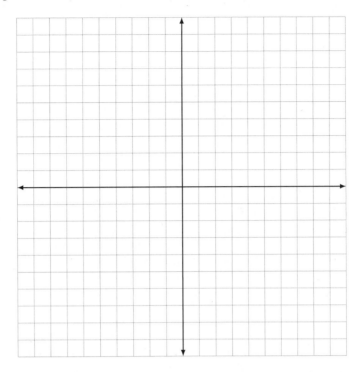

6 If a line L_1 has a slope of $\frac{4}{7}$, then find the slope of a line that is perpendicular to L_1. Also, find the slope of a line that is parallel to L_1. Hence, find the product of the two slopes.

7 Find a line equation that contains $(3,4)$ and perpendicular to the line $x - 3y = 12$.

8 The line $y = ax + b$ is perpendicular to the line $y + 2x = 8$ and meets the line $y + x + 3 = 0$ on the y-axis. Find the value of a and b.

9 Graph the line $y = |2 - x| + 2$ on the following coordinate plane.

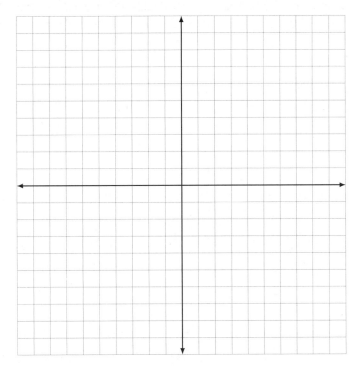

10 Sketch(or shade) the graph of the linear inequality $2x + y > 1$.

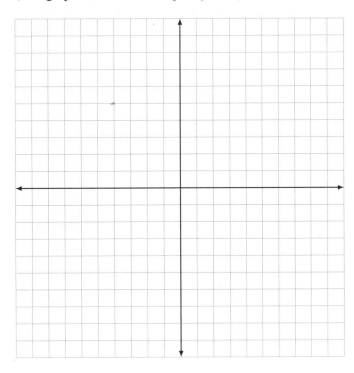

11 Suppose there is a new calendar known as 'metric calendar,' which is equivalent to the usual calendar of 365 days. In other words, 1 metric year equals 365 usual days. Since 1 metric year equals 10 metric months, 1 metric month equals 10 metric weeks, and 1 metric week equals 10 metric days, 1000 metric days equal 365 usual days. How many usual days are there in 458 metric days?

12 The ratio of A to B can be rewritten as $A : B$ or A/B. If we have to use it as a number, we use the latter form. Otherwise, we write it using the colon method. Assuming that x and y are real numbers, if the ratio of $2x - y$ to $x + y$ is $2 : 3$, find the ratio $x : y$.

13 If a stack of 8 quarters is exactly one-half inch high, how many quarters will be needed to make a stack one foot high? (Hint : 1 foot = 12 inches.)

14 If y^2 varies inversely as x^3, and $y = 3$ when $x = 2$, find the value of y when $x = 9$, assuming that $y > 0$.

1 (a) $f(3) - f(-2) = 15$ (b) Since the slope is 3, so $\dfrac{f(3) - f(-2)}{3 - (-2)} = 3$, implying $f(3) - f(-2) = 15$.

2 The 4th Quadrant.

3 $b = 2$

4 The product of the slope and y-intercept is $-\dfrac{8}{3}$.

5

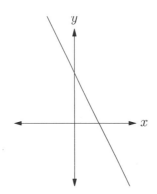

6 The product equals -1.

7 The line equation is $y = -3x + 13$.

8 $a = \dfrac{1}{2}$ and $b = -3$.

9

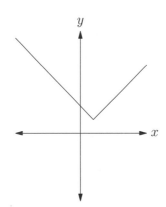

10

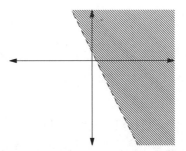

11 167.17 days

12 $x : y = 5 : 4$

13 192 quarters

14 $y = \dfrac{2\sqrt{2}}{9}$

저자가 알려주는 <The Essential Guide to Algebra 2> 효율적으로 사용하기 위한 Tip!

Q. 인터넷 강의 혹은 현장 강의를 통해 학습한다면?

첫째, 인터넷 강의 또는 현장 강의를 통하여, 꼼꼼히 필기를 하며 개념을 훑어 학습합니다.

둘째, 완강을 한 이후, 1주일 정도의 복습 시간을 통하여, 해당 문제들에서 중심이 되는 내용들을 기억해내며 다시 한번 정리합니다.

셋째, 교재 필기와 자신이 생각한 노트 등을 잘 정리해두었다가, 학교에서 Algebra 2를 수강하는 시점에, 시험 또는 퀴즈 직전에 다시 한번 내용을 잘 볼 수 있도록 정리해 둡니다.

Q. 혼자 학습한다면?

첫째, 문제를 들어가기 전에 교재에 적힌 개념 설명을 읽어보고, 본문의 문제를 풀기 시작합니다.

둘째, Solution Manual에 있는 내용을 보며, 자신이 생각한 답과 비교하며, 정리해둡니다.

셋째, Skill Practice에 나와 있는 내용의 문제를 풀어보며, 자신이 알고 있는 내용을 탄탄히 만들도록 합니다.

Topic 3

System of Equations and Inequalities

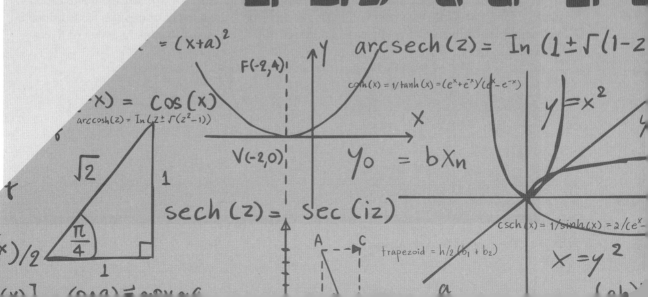

3.1 System of Linear Equations

Given a system of linear equations

$$\begin{cases} ax + by = c \\ dx + ey = f \end{cases}$$

we can classify the system as <u>independent</u>, <u>dependent</u>, or <u>inconsistent</u>.

- $\begin{cases} x + y = 3 \\ x - 2y = 1 \end{cases}$ is consistent and independent.

- $\begin{cases} x + 3y = 4 \\ 9y + 3x = 12 \end{cases}$ is consistent and dependent.

- $\begin{cases} x + 2y = 1 \\ 2x + 4y = 1 \end{cases}$ is inconsistent.

Example

How do we know whether $\begin{cases} x + 2y = 1 \\ 2x + 4y = 1 \end{cases}$ is inconsistent?

Solution

The quickest method of figuring out whether the system of consistent, we look at the ratio of coefficients of respective variables. Let's look at the ratio of coefficients of x and y. From top to bottom, we get $\frac{1}{2}$ for both variables. This indicates two possibilities. We either get a consistent, dependent system so that there are infinitely many solutions to the system, or an inconsistent system so that there is no solution whatsoever. This is directly determined by the ratio of the constants. From top to bottom, we get $\frac{1}{1}$, which is different from $\frac{1}{2}$. This means that these two lines are parallel, having no common solution between the two equations. What if the ratio of constants were $\frac{1}{2}$? We may conclude in this case that two lines are equivalent, meaning that the system is consistent and dependent.

1 You and your business partner are mailing advertising flyers to your customers. You address 6 flyers each minute and have already done 80. Your partner addresses 4 flyers each minute and has already done 100. Write the system of linear equations on t, the number of minutes passed since you addressed 80 flyers and your partner addressed 100 flyers, and f, the total number of flyers addressed so far.

2 Jane is going on a vacation and leaving her dog in a kennel. Kennel X charges $25 per day, which includes one-time grooming treatment. Kennel Y charges $20 per day and one-time fee of $30 for grooming. (Assume that kennel X and Y offer only one grooming service, and her dog remains equally happy in both places.)

(a) Write a system of equations to represent the cost c, in dollars, for d days that your dog will stay at the kennel.

(b) If your vacation is 7 days long, which kennel should you choose? Justify your reasoning focused on economic perspectives.

3 Which of the system of equations is inconsistent?

(A) $\begin{cases} x+y=4 \\ x-y=3 \end{cases}$

(B) $\begin{cases} 6x+3y=12 \\ 2y+4x=4 \end{cases}$

(C) $\begin{cases} 2y-x=5 \\ 4y-2x=10 \end{cases}$

(D) $\begin{cases} -3x+y=4 \\ 2y+6x=8 \end{cases}$

4 Jimmy and Robert are both knitting scarves for charity. In average, Jimmy knits 8 rows each minute and already has knitted 10 rows. Likewise, Robert knits 5 rows each minute and has already knitted 19 rows. When will both have knitted the same number of rows?

(A) 2.6 minutes
(B) 3 minutes
(C) 9.7 minutes
(D) 34 minutes

5 The solution pair (x, y) to the system of equations is the point of intersection (x, y) between the two graphs. Graph the following system of equations in the diagram below.

$$\begin{cases} x + y = 3 \\ 2x - 2y = 4 \end{cases}$$

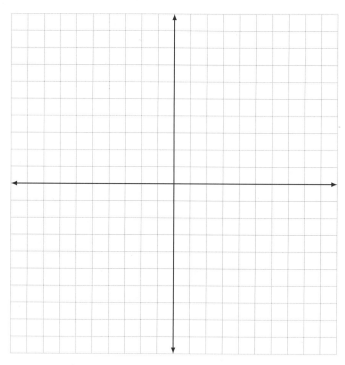

3.2   Two Methods : Substitution and Elimination

This is a quick review section for solving the system of equations. The first technique is called substitution. In other words, we can solve the system of equations by substitution. Let's look at the following example.

Example

Solve $\begin{cases} y = x + 1 \\ 2x + y = 7 \end{cases}$

Solution
Substitute $y = x + 1$ into the second equation such that $2x + (x + 1) = 7$. Hence, $3x = 6$ so $x = 2$ and $y = 3$.

On the other hand, we can eliminate the common variables (or reduce the number of variables) in the system of linear equations, the technique of which we learned in Algebra I.

Example

Solve $\begin{cases} x - y = 2 \\ 2x + y = 7 \end{cases}$

Solution
Adding the two equations vertically results in $3x = 9$. Therefore, $x = 3$ and $y = 1$.

[6] Suppose you bought eight oranges and one grapefruit for a total of $4.60. Later that day, you bought six oranges and three grapefruits for a total of $4.80. What is the price of each type of fruit?

7 There are a total of 15 apartments in two buildings. The difference of two times the number of apartments in the first building and three times the number of apartments in the second building is 5. Assume there are more apartments in the first building than in the second building.

(a) Write the system of equations to model the given situation with f and s, where f is the number of apartments in the first building and s the number of apartments in the second building.

(b) How many apartments in each building?

8 You bought books at a book store for a fixed price of $15.49 each. You could have bought them at an online store for $13.99 each plus $6 for one-time shipping fee. In other words, the online store charges a fixed price of $6 for delivery, no matter how many books you buy in the online store. How many books can you buy for the same amount at the two stores?

9 Last year, a baseball team paid $20 per bat and $12 per glove, spending a total of $1,040. This year, the prices went up to $25 per bat and $16 per glove. The team spent $1,350 to purchase the same amount of equipment as last year. How many bats and gloves did the team purchase each year?

10 If the perimeter of the square below is 72 units, what are the values of x and y?

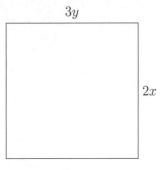

In the previous topic, we learned how to plot linear inequality. This is another review of two-variable inequalities. We simply find and color the region that is common to both inequalities.

Example

Solve $\begin{cases} y < x+2 \\ y \leq -x+1 \end{cases}$.

Solution

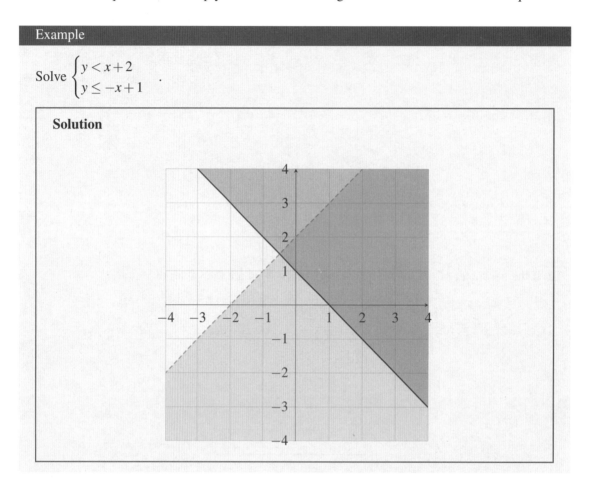

11 Suppose you are buying two kinds of scratch papers for school. A spiral notebook costs $2, and a three-ring notebook costs $5. You must have at least 6 notebooks, whereas the cost of the notebooks can be no more than $20. Write a system of inequalities. (Do not solve the system of inequalities).

In the following example, color the region common to both inequalities. In part (b), label the grid properly to fit the shaded region within the figure.

12 Solve the following system of linear inequalities. Assume $x \geq 0$ and $y \geq 0$.

(a) $\begin{cases} 2x + 4y \leq 12 \\ x + y \leq 5 \end{cases}$

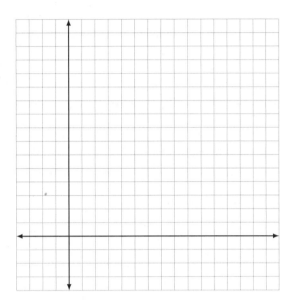

(b) $\begin{cases} 3x + 5y \leq 30 \\ x + y \geq 8 \end{cases}$

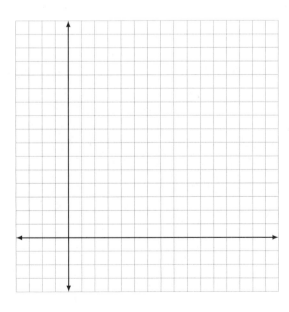

3.4 Linear Programming

We change sentences into mathematical statements in order to find the optimal solution.

Example

A computer manufacturer makes 30 to 50 monitors a day and 40 to 70 keyboards a day. At most, he can produce 100 units a day. The profit on a monitor is \$70 and the profit on a keyboard is \$50. How many monitors and keyboards should he make everyday to maximize the profit? How much is the profit?

Solution

Let's write down inequalities by labelling variables. Here, inequalities are known as *constraints*.

$$x = \text{the number of monitors produced in a day}$$
$$y = \text{the number of keyboards produced in a day}$$

All possible constraints can be written as

$$30 \leq x \leq 50$$
$$40 \leq y \leq 70$$
$$x + y \leq 100$$

Graph all three inequalities to find the feasible region(or overlapping region). Find the vertices of the region and substitute them into the equation that we should optimize. We choose the largest to find out the maximum profit.

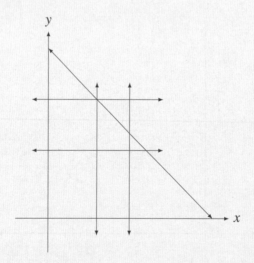

The corner points are $(30, 70)$, $(30, 40)$, $(50, 40)$, $(50, 50)$. The profit expression we should maximize is $70x + 50y$. Substituting all four points, we get $(50, 50)$ producing the maximum profit. Hence, the manufacturer should produce 50 monitors and 50 keyboards a day, making $6,000$ dollars per day.

13 The area of a parking lot is 600 square meters. A car requires 6 square meters. A bus requires 30 square meters. The attendant can handle only 60 vehicles. Assume that this attendant is great at parking so that there is no empty space between the vehicles whatsoever. If a car is charged $2.50 and a bus $7.50, how many of each should be accepted to maximize income? (Use the following xy-plane to sketch the feasible region.)

(a) Set up the correct constraints.

(b) Draw the feasible region.

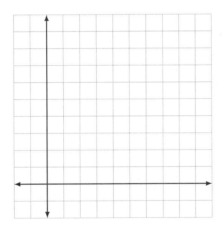

(c) Evaluate the maximum income.

14 The B & W Leather Company wants to add handmade belts and wallets to its product line. Each belt nets the company $18 in profit, and each wallet nets $12. Both belts and wallets require cutting and sewing. Belts require 2 hours of cutting time and 6 hours of sewing time. Wallets require 3 hours of cutting time and 3 hours of sewing time. If the cutting machine is available 12 hours a week and the sewing machine is available 18 hours per week, what ratio of belts and wallets will produce the most profit within the constraints? (Assume that the sizes of belts and wallets come out in half unit. In other words, it is possible to produce 2.5 belts or 1.5 wallets.)

(a) Set up the correct constraints.

(b) Draw the feasible region.

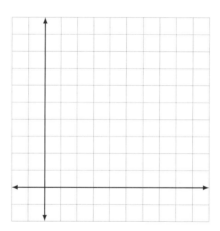

(c) Evaluate the maximum profit.

15 The Cruiser Bicycle Company makes two styles of bicycles: the Traveler, which sells for $300, and the Tourister, which sells $600. Each bicycle has the same frame and tires, but the assembly and painting time required for the Traveler is only 1 hour, while it is 3 hours for the Tourister. There are 300 frames and 360 hours of labor available for production. How many bicycles of each model should be produced to maximize revenue, given the constraints?

(a) Set up the correct constraints.

(b) Draw the feasible region.

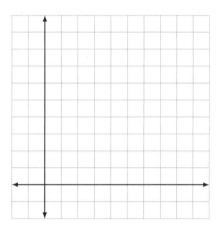

(c) Evaluate the maximum revenue.

3.5 System of Equations in 3 Variables

First, the three dimensional points we study in Algebra 2 usually form a plane - a flat surface that infinitely extends - whose equation is given by a linear equation of three variables.

Example

$x+y+z=2$ is the plane that passes through $(1,1,0)$, $(1,0,1)$, and $(0,1,1)$.

16 Which of the following points is on the graph of $2x+3y+z=12$?

(A) $(1,2,3)$
(B) $(2,1,3)$
(C) $(3,0,5)$
(D) $(0,3,3)$

The following example will show how to plot (x,y,z). Each entry (a number) in respective variable indicates the amount of movement from the origin to respective direction. Let's have a look at the following example.

Example

Explain how to plot $(3,2,5)$ in the xyz plane.

Solution
Move three units to the x-axis from the Origin. Then, move two units to the y-axis. Lastly, move five units to the z-axis direction. This is the point that corresponds to $(3,2,5)$.

17 Find the distance between the origin $(0,0,0)$ and the point $(1,3,4)$.

In order to draw a line in the plane, we need *two distinct points*. Likewise, in order to draw a plane, we need *three distinct points*. The idea behind this phenomenon can be easily explained by a triangle that sits inside the plane. Imagine yourself drawing a triangle on a clean sheet of paper. As you could guess from there, there is only one plane that contains that triangle. Read the following example.

Example

Graph $2x + 3y + 4z = 1$.

> **Solution**
> The fastest way to find three points satisfying this equation is to find x-intercept, y-intercept, and z-intercept.
>
> | x | y | z |
> |---|---|---|
> | $\frac{1}{2}$ | 0 | 0 |
> | 0 | $\frac{1}{3}$ | 0 |
> | 0 | 0 | $\frac{1}{4}$ |
>
> Connect these three points to produce triangle. There is only one plane that contains this triangle. This plane is the plane we want to find.

A system of three linear equations with three variables deals with three planes. That being written, there are only three possible cases to cover.

(1) No solution : Two planes meet, while all three do not meet at the same point or same line.

(2) Infinitely many solutions : Three planes meet at a line. The number of points of intersection, in this case, is not finite.

(3) Exactly one solution : There is a unique point of intersection between three planes. Recall a conventional xyz plane. The origin is a point of intersection between xy plane, yz plane, and xz plane.

18 Which of the following must be the possible intersection points of three distinct planes?

(A) Single point only
(B) Line only
(C) Point and Line
(D) Plane

There are two ways to solve the system of equations in three variables : either substitution or elimination. The key idea of solving the system of equations is to reduce the number of variables down all the way down to one.

19 Solve $\begin{cases} x+y+z=1 \\ 2x+y+z=5 \\ y-z=5 \end{cases}$ by substitution.

20 Solve $\begin{cases} x-y+z=1 \\ 2x+y-z=-1 \\ x-2y+z=3 \end{cases}$ by elimination.

3.6 Matrices

Matrix is an array of numbers in rows and columns. The reason why we learn it in this topic is because matrix helps us write a system of equations in different methods.

- The order of matrix: Suppose there are m rows and n column. Then, we say a matrix has the dimension of $m \times n$, read as "m by n." For example, a 2×3 matrix A can be specified as

$$\begin{bmatrix} x_{11} & x_{12} & x_{13} \\ x_{21} & x_{22} & x_{23} \end{bmatrix} = A_{2 \times 3}$$

- If two matrices have same entries with equal dimensions, we say they are equal.

$$\begin{bmatrix} x_{11} & x_{12} \\ x_{21} & x_{22} \end{bmatrix} = \begin{bmatrix} w_{11} & w_{12} \\ w_{21} & w_{22} \end{bmatrix} \implies x_{11} = w_{11}, x_{12} = w_{12}, x_{21} = w_{21}, x_{22} = w_{22}$$

Example

A drinks stall sold 160 cups of coffee, 125 cups of tea and 210 glasses of soft drinks on Monday. Similarly, it sold 145 cups of coffee, 130 cups of tea and 275 glasses of soft drinks on Tuesday. Lastly, it sold 120 cups of coffee, 155 cups of tea and 325 glasses of soft drinks on Wednesday. Design a matrix to represent this information, labeling the rows and columns. State the order of matrix.

Solution

Let the first column be the sales of coffee, second be the sales of tea, and the last column be the sales of soft drink. Then, the dimension of matrix is the product of 3×3 matrix and 3×1 matrix.

$$\begin{bmatrix} 160 & 125 & 210 \\ 145 & 130 & 275 \\ 120 & 155 & 325 \end{bmatrix} \begin{bmatrix} c \\ t \\ s \end{bmatrix}$$

21 Which of the following is the correct dimension of the following matrix?

$$\begin{bmatrix} -1 & 3 & 4 & 12 \\ 2 & 3 & 4 & 10 \\ -5 & -3 & 3 & 2 \end{bmatrix}$$

(A) 3×4 (B) 4×3 (C) 4×4 (D) 3×3

Adding or subtracting two matrices of equal dimension is simple - add or subtract entries of the same position. However, it is *impossible* to add or subtract matrices of different dimensions.

$$\begin{bmatrix} x_{11} & x_{12} & x_{13} \\ x_{21} & x_{22} & x_{23} \end{bmatrix} \pm \begin{bmatrix} y_{11} & y_{12} & y_{13} \\ y_{21} & y_{22} & y_{23} \end{bmatrix} = \begin{bmatrix} x_{11} \pm y_{11} & x_{12} \pm y_{12} & x_{13} \pm y_{13} \\ x_{21} \pm y_{21} & x_{22} \pm y_{22} & x_{23} \pm y_{23} \end{bmatrix}$$

Furthermore, the scalar multiple is the product of a matrix and a constant, which simply multiply that number to every entry of the given matrix. Have a look at the following expression to see what it means to multiply a constant to every entry of the given matrix.

$$k \begin{bmatrix} y_{11} & y_{12} & y_{13} \\ y_{21} & y_{22} & y_{23} \end{bmatrix} = \begin{bmatrix} ky_{11} & ky_{12} & ky_{13} \\ ky_{21} & ky_{22} & ky_{23} \end{bmatrix}$$

Example

If $A = \begin{bmatrix} 0 & 0 \\ 1 & 1 \end{bmatrix}, B = \begin{bmatrix} 1 & 0 \\ 1 & 1 \end{bmatrix}$, then find the sum of all entries for $2A + B$.

Solution

$2A + B = \begin{bmatrix} 0 & 0 \\ 2 & 2 \end{bmatrix} + \begin{bmatrix} 1 & 0 \\ 1 & 1 \end{bmatrix} = \begin{bmatrix} 1 & 0 \\ 3 & 3 \end{bmatrix}$. Hence, the sum of entries is $1 + 0 + 3 + 3 = 7$.

22 The sum of entries of $2 \begin{bmatrix} 3 & 4 \\ 2 & -1 \end{bmatrix} - 3 \begin{bmatrix} 1 & -1 \\ -3 & 2 \end{bmatrix}$ is equal to

(A) 13
(B) 15
(C) 19
(D) 21

3.7 Multiplication of Matrix

Assume that $m \neq n \neq p$. Given $A_{m \times n}$ and $B_{n \times p}$, the dimension of AB is $m \times p$. The matrix multiplication is obtainable since the number of columns of the leftmost matrix equals the number of rows in the rightmost matrix. On the other hand, BA does not exist because $p \neq m$. In other words, the number of columns in the leftmost matrix does not match with that of rows in the rightmost matrix.

How do we find out the entry of the new matrix AB? New entry is found by a process called row-column multiplication, which is also known as a dot-product. It is called "row-column" multiplication because it chooses a row from the leftmost matrix and a column from the rightmost matrix to attain the dot product.

$$\begin{bmatrix} x_{11} & x_{12} & x_{13} \\ \cdots & \cdots & \cdots \end{bmatrix} \times \begin{bmatrix} y_{11} & \cdots \\ y_{21} & \cdots \\ y_{31} & \cdots \end{bmatrix} = \begin{bmatrix} z_{11} & \cdots \\ \cdots & \cdots \end{bmatrix}$$

Here, z_{11} is the entry of the first row and first column of the new matrix. As the name of the row-column multiplication suggests, choose the first row of the first matrix and the first column of the second matrix to get dot product, i.e., $z_{11} = x_{11}y_{11} + x_{12}y_{21} + x_{13}y_{31}$.

Example

Find $\begin{bmatrix} 2 & 1 \\ 3 & -2 \end{bmatrix} \times \begin{bmatrix} 1 & 0 \\ 3 & 2 \end{bmatrix}$.

Solution

If we use the multiplication method illustrated above, $\begin{bmatrix} 5 & 2 \\ -3 & -4 \end{bmatrix}$. In particular, each entry is easily computed as

$$5 = 2 \cdot 1 + 1 \cdot 3$$
$$2 = 2 \cdot 0 + 1 \cdot 2$$
$$-3 = 3 \cdot 1 + (-2) \cdot 3$$
$$-4 = 3 \cdot 0 + (-2) \cdot 2$$

23 Evaluate $\begin{bmatrix} 1 & 0 \\ 0 & -1 \end{bmatrix} \times \begin{bmatrix} 2 & 3 \\ -1 & 2 \end{bmatrix}$

From now on, we would like to focus on square matrices, especially 2×2 ones.[1] Identity matrix is written as $I_{2\times 2}$ (or sometimes, $E_{2\times 2}$), where it refers to $\begin{bmatrix} 1 & 0 \\ 0 & 1 \end{bmatrix}$. In this matrix, all the diagonal entries are equal to 1. In short, if you multiply the identity matrix to different square matrix, you get that matrix again.

$$A_{2\times 2} \cdot I_{2\times 2} = I_{2\times 2} \cdot A_{2\times 2} = A_{2\times 2}$$

Now, we can talk about the inverse matrix can be found in the following method. If the given matrix is a 2×2 matrix, $A_{2\times 2} = \begin{bmatrix} a & b \\ c & d \end{bmatrix}$, then the inverse matrix A^{-1} is computed as

$$A^{-1} = \frac{1}{ad-bc} \begin{bmatrix} d & -b \\ -c & a \end{bmatrix}$$

Obviously, the denominator expression *cannot* be 0. This number $ad - bc$ is known as the determinant, $det(A)$. If it is 0, then A is known to be singular matrix, where A^{-1} does not exist. Otherwise, the inverse matrix of A surely exists. Why do you think we call it as an inverse matrix? Think about multiplicative inverse of a number. If we multiply x by x^{-1}, then we get 1. Same logic is applied to the world of matrices, i.e., the product of A and A^{-1} results in the identity matrix, i.e., $A_{2\times 2} \cdot A_{2\times 2}^{-1} = A_{2\times 2}^{-1} \cdot A_{2\times 2} = I_{2\times 2}$.

24

(a) Find the inverse, A^{-1}, of a matrix $A = \begin{bmatrix} 2 & 1 \\ -1 & 3 \end{bmatrix}$.

(b) Find the value of a natural number n if all of the entries of the inverse of $M = \begin{bmatrix} 2n & -7 \\ -1 & n \end{bmatrix}$ are natural numbers.

---

[1]Normally, if $m \neq n$, then we say $A_{m\times n}$ is a rectangular matrix. However, if the number of rows and columns coincides with each other, we say that the matrix is a square matrix.

Now, we are ready to connect the system of equations with matrices. Let's suppose we have a system of equations $ax + by = m$ and $cx + dy = n$. We could easily convert this form into the matrix equation, i.e.,

$$\begin{bmatrix} a & b \\ c & d \end{bmatrix}_{2\times 2} \times \begin{bmatrix} x \\ y \end{bmatrix}_{2\times 1} = \begin{bmatrix} m \\ n \end{bmatrix}_{2\times 1}$$

How do we solve the equation using matrix method? Follow the steps[2] laid out below to solve the following question in the next example.

$$AB = C$$
$$A^{-1}(AB) = A^{-1}C$$
$$(A^{-1}A)B = A^{-1}C$$
$$IB = A^{-1}C$$
$$B = A^{-1}C$$

25 Solve the following matrix equation to get the values of x and y.

$$\begin{bmatrix} 3 & -1 \\ -2 & 3 \end{bmatrix} \begin{bmatrix} x \\ y \end{bmatrix} = \begin{bmatrix} 4 \\ 2 \end{bmatrix}$$

26 Find the determinant of the following matrix $\begin{bmatrix} 2 & 1 \\ 4 & 2 \end{bmatrix}$.

---

[2]Matrix multiplication is not *commutative*, meaning that the order of multiplication matters. However, it is *associative*, meaning that the grouping order does not matter. In other words, if we multiply a proper matrix at the front of the both sides of the equation, we could make identity matrix in the left-side of the equation to solve for unknowns.

1 If the sum of two numbers is 4 and the sum of their squares minus three times their product is 76, find the numbers.

2 If $(1, p)$ is a solution of simultaneous equations,

$$\begin{cases} 12x^2 - 5y^2 = 7 \\ 2p^2x - 5y = 7 \end{cases}$$

evaluate the value of p and the other solution.

3 Solve the following system of equations.

$$\begin{cases} x+y=9 \\ xy=8 \end{cases}$$

4 Find the coordinates of the points where the line meets the curve.

$$\begin{cases} 2x+3y=-1 \\ x(x-y)=2 \end{cases}$$

5 Given that $A = \begin{bmatrix} 1 & 1 \\ 0 & 0 \end{bmatrix}$, find A^{23}.

6 If $A = \begin{bmatrix} 0 & 1 \\ 1 & 0 \end{bmatrix}, B = \begin{bmatrix} 1 & 1 \\ 0 & 1 \end{bmatrix}$, find $A^2B - A$. Use the following equivalent forms of $A^2B - A$, which is given by[3]

$$A^2B - A = A(AB) - AI$$
$$= A(AB - I)$$

---

[3]Matrix multiplication allows *distributive property*. As long as we use $I_{2\times2}$ wisely, we can come up with a fancy way of factorization to easily compute the matrix expressions.

7 Given two 2×2 matrices M, N satisfying

$$M = \begin{bmatrix} 2 & -4 \\ -1 & 2 \end{bmatrix} \qquad N = \begin{bmatrix} 1 & 2 \\ 2 & 4 \end{bmatrix}$$

find $\dfrac{1}{3}MN - NM$.

8 Given $A = \begin{bmatrix} 1 & 1 \\ 1 & 0 \end{bmatrix}, B = \begin{bmatrix} 1 & 2 \\ 3 & 4 \end{bmatrix}$, if $2A + X = AB$, then find a matrix $X_{2 \times 2}$.

9 Find the sum of all entries for A^{-1} if $A = \begin{bmatrix} 1 & -2 \\ 0 & 1 \end{bmatrix}$.

10 Find the sum of entries for $X_{2\times2}$ if $A = \begin{bmatrix} 1 & 2 \\ 2 & 5 \end{bmatrix}, B = \begin{bmatrix} 2 & -3 \\ 1 & -2 \end{bmatrix}$ and $A \cdot X_{2\times2} = B$.

1. 6 and -2.

2. $p = -1$ and the other solution is $\left(-\dfrac{3}{2}, -2\right)$.

3. $(x, y) = (1, 8)$ or $(8, 1)$.

4. $(x, y) = \left(-\dfrac{6}{5}, \dfrac{7}{15}\right), (1, -1)$.

5. $A^{23} = \begin{bmatrix} 1 & 1 \\ 0 & 0 \end{bmatrix}$.

6. $A^2 B - A = \begin{bmatrix} 1 & 0 \\ -1 & 1 \end{bmatrix}$.

7. $\dfrac{1}{3} MN - NM = \begin{bmatrix} -2 & -4 \\ 1 & 2 \end{bmatrix}$.

8. $X = \begin{bmatrix} 2 & 4 \\ -1 & 2 \end{bmatrix}$.

9. The sum of entries of A^{-1} is 4.

10. The sum of entries of X is -2.

유하림 저자의 기출간 교재 LIST

- 몰입공부
- The Essential Workbook for SAT Math Level 2
- The Essential Guide to IGCSE : Additional Math (인터넷 강의 전용)
- The Essential Guide to Prealgebra
- The Essential Guide to Algebra 1
- The Essential Guide to Geometry
- The Essential Guide to Precalculus
- The Essential Guide to Competition Math : Fundamentals
- The Essential Guide to Competition Math : Number Theory
- The Essential Guide to Competition Math : Counting and Probability

Topic 4

Quadratic Equations and Factoring

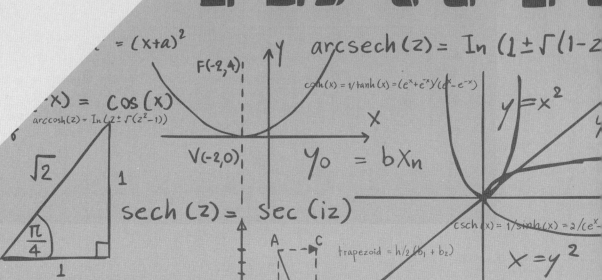

4.1 Quadratic Function

The graph of a quadratic function is known as parabola that has a vertex at either the lowest point or the highest point. It has the lowest point in the form of $y = x^2$. On the other hand, it has the highest point in the form of $y = -x^2$. The following diagram is the graph of $y = f(x) = x^2$.

As one can see from the following graph, the graph has two disinct x-values for one y-value. This calls out for line symmetries. You should hear your inner voice, "The midpoint!" In fact, this function is called "even" function because the graph is symmetric about the y-axis. In other words, $f(x) = f(-x)$. So far, we learned two functions that are even : $y = |x|$ and $y = x^2$.

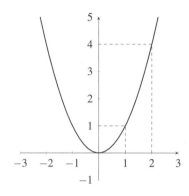

$\boxed{1}$ Graph the following function.

(a) $y = \dfrac{1}{2}x^2$

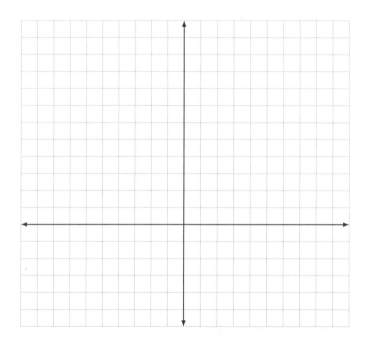

The coefficient of x^2 determines whether the graph is concave up or down. If it is positive, then the graph is concave up. Otherwise, the graph is concave down.

(b) $y = -x^2$

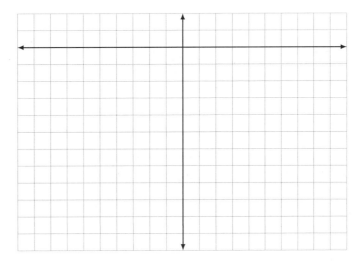

The graph of $y = k(x^2)$ and $y = (kx)^2$ may look alike, but the process behind the transformation is completely different. In part (a), we looked at the graph is vertically shrunk. Nevertheless, the following example is all about dilating the width, yet maintaining the height.

(c) $y = \left(\dfrac{x}{2}\right)^2$

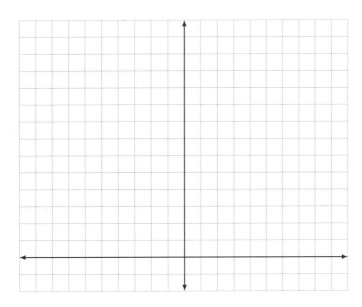

Identify the vertex, axis of symmetry, the maximum or minimum value, and the domain and the range of each function $f(x) = 0.3(x-1)^2 + 4$.

Solution

- The vertex is $(1,4)$.

- The axis of symmetry is $x = 1$.

- The minimum value is $y = 4$.

- The domain is $\mathbb{R}$.

- The range is $[4, \infty]$.

2 Identify the vertex, axis of symmetry, the maximum value, and the domain and the range of each function.

$$y = -2(x+3)^2 + 2$$

(a) Vertex   (b) Axis of Symmetry   (c) Maximum value   (d) Range

3 Describe how to transform the parent function $y = x^2$ to the graph of each of the following functions below. If one of the transformation is not applied in the given function, write down "not applied."

(a) $y = -(x+1)^2 - 3$

- Dilation :

- Reflection :

- Translation :

(b) $y = \dfrac{1}{2}(x-1)^2 - 1$

- Dilation :

- Reflection :

- Translation :

Given a standard form of a quadratic function $y = ax^2 + bx + c$, we call

- $a(x-h)^2 + k$: a vertex form

- $a(x-m)(x-n)$: an intercept form

4 Write down the equation of each parabola in vertex form.

(a) vertex $(3, 1)$ and point $(1, 3)$

(b) vertex $\left(\frac{1}{2}, 1\right)$ and point $(2, -1)$

5 The amount of cloth used to produce four curtains is given by the function $A(x) = -4x^2 + 40x$ where x is the width of one curtain in inches and A is the total area in square inches. Find the width that maximizes the area of the curtains. What is the maximum area?

4.2 Standard Form of a Quadratic Function

The function $y = ax^2 + bx + c$ is in the standard form such that

- a determines the concavity of the graph. If $a > 0$, then the graph is concave up. If $a < 0$, then the graph is concave down.

- b determines the sum of roots. In fact, $-\dfrac{b}{a}$ is the sum of roots.

- c determines the y-intercept.

Example

Identify the vertex, the axis of symmetry, the maximum or minimum value, and the range of the following parabola $y = x^2 - 4x + 3$.

Solution
$y = x^2 - 4x + 3 = (x^2 - 4x + 4 - 4) + 3 = (x^2 - 4x + 4) - 4 + 3 = (x-2)^2 - 1$. Hence, the vertex is $(2, -1)$, the axis of symmetry $x = 2$, the minimum value of $y = -1$ and the range is $[-1, \infty)$.

$\boxed{6}$ Identify the vertex, the axis of symmetry, the maximum or minimum value, and the range of the following parabolas.

(a) $y = 3x^2 + 12x + 15$

(b) $y = -2x^2 + 4x$

(c) $y = 2x^2 - x + 1$

7 Write the following quadratic functions in vertex form.

(a) $y = x^2 - 6x + 9$

(b) $y = x^2 - 4x$

(c) $y = 2x^2 - 10x + 2$

(d) $y = -3x^2 + 2x + 1$

8 A small independent publishing company determines the profit P for publishing n new books is

$$P(n) = -0.02n^2 + 3.40n - 16$$

where P is the profit in thousands of dollars and n is in thousands of units.

(a) How many books should the company publish to maximize the profit?

(b) What is the maximum profit?

9 Given a quadratic function $y = ax^2 - 10x + b$, if $(5,2)$ is the vertex of the function's graph, find the unknown constants a and b.

4.3 Factoring Quadratic Expressions

Given $x^2 + bx + c = 0$ where r and s are its roots, then we can factor it into $(x-r)(x-s)$. In short, $x^2 - (r+s)x + rs = 0$. In other words, we are looking for two integer divisors of c whose sum equals b.

Example

Factor $x^2 + 11x + 28$.

> **Solution**
> Let's investigate 28 closely. Since $28 = 1 \cdot 28$, $2 \cdot 14$, $4 \cdot 7$, $(-1) \cdot (-28)$, $(-2) \cdot (-14)$, $(-4) \cdot (-7)$, we will check whether the sum of the integer pairs is 11.
>
> - $1 + 28 \neq 11$
>
> - $2 + 14 \neq 11$
>
> - $4 + 7 = 11$
>
> - $-1 - 28 \neq 11$
>
> - $-2 - 14 \neq 11$
>
> - $-4 - 7 \neq 11$
>
> Hence, $x^2 + 11x + 28 = (x+4)(x+7)$.

---

10 Factor each expression into the product of linear terms with integer coefficients. If it cannot be factored, then write "irreducible."

(a) $x^2 - 10x + 21$

(b) $x^2 - 12x + 32$

(c) $-x^2 + 12x - 35$

(d) $-y^2 - 3y + 54$

(e) $x^2 + 7x - 60$

(f) $x^2 - 8x + 15$

11 Factor each expression into the product of linear terms with integer coefficients. If it cannot be factored, then write "irreducible."

(a) $5x^2 - 17x + 6$

(e) $3x^2 + 10x + 8$

(b) $2x^2 - 9x + 5$

(f) $x^2 + 12x + 36$

(c) $9x^2 - 6x + 1$

(g) $4x^2 + 12x + 9$

(d) $x^2 - 49$

(h) $2x^2 - 50$

4.4 Completing the Square

From $ax^2 + bx + c = 0$, it is convenient for us to change it into a vertex form $a(x-h)^2 + k$ to analyze the behavior of quadratic function. The process of converting the standard form into the vertex form is called "completing the square."

Completing the Square

Complete the squares $x^2 + 14x +$ _____.

Solution
Divide 14 by 2 and square the result, i.e., 49. Hence, $x^2 + 14x + 49 = (x+7)^2$.

12 Complete the square by filling the last constant to the quadratic expressions.

(a) $x^2 - 30x+$

(d) $2x^2 + 4x+$

(b) $x^2 + 5x+$

(e) $20x^2 + 10x+$

(c) $x^2 - \dfrac{1}{3}x+$

(f) $4x^2 - 10x+$

13 Find the positive value of k that makes the left side of each equation a perfect square trinomial.

(a) $x^2 + kx + 144$

(b) $36x^2 + kx + 1$

(c) $x^2 - kx + 25$

(d) $16x^2 - kx + 9$

14 The quadratic function $d(t) = -t^2 + 6t + 23$ models the depth of water in a flood channel after a rainstorm. The time in hours after it stops raining is t and $d(t)$ is the depth of the water in feet. Find the maximum depth of water and when it occurs.

4.5   Quadratic Formula

Given $ax^2 + bx + c = 0$, we can deduce the quadratic formula to directly find the roots of the equation.

$$ax^2 + bx + c = 0 \tag{4.1}$$

$$ax^2 + bx = -c \tag{4.2}$$

$$x^2 + \frac{b}{a}x = -\frac{c}{a} \tag{4.3}$$

$$x^2 + \frac{b}{a}x + \left(\frac{b}{2a}\right)^2 = -\frac{c}{a} + \left(\frac{b}{2a}\right)^2 \tag{4.4}$$

$$\left(x + \frac{b}{2a}\right)^2 = -\frac{c}{a} + \frac{b^2}{4a^2} \tag{4.5}$$

$$\left(x + \frac{b}{2a}\right)^2 = \frac{-4ac}{4a^2} + \frac{b^2}{4a^2} \tag{4.6}$$

$$\left(x + \frac{b}{2a}\right)^2 = \frac{b^2 - 4ac}{4a^2} \tag{4.7}$$

$$x + \frac{b}{2a} = \pm\sqrt{\frac{b^2 - 4ac}{4a^2}} \tag{4.8}$$

$$x + \frac{b}{2a} = \pm\frac{\sqrt{b^2 - 4ac}}{\sqrt{4a^2}} \tag{4.9}$$

$$x + \frac{b}{2a} = \pm\frac{\sqrt{b^2 - 4ac}}{2a} \tag{4.10}$$

$$x = -\frac{b}{2a} \pm \frac{\sqrt{b^2 - 4ac}}{2a} \tag{4.11}$$

$$x = \frac{-b \pm \sqrt{b^2 - 4ac}}{2a} \tag{4.12}$$

15 Use the quadratic formula to solve each equation.

(a) $x^2 = 4x + 1$ (b) $2x^2 - 10x + 6 = 0$

The discriminant $b^2 - 4ac$ is the expression inside the square root of the quadratic formula. The expression is called *discriminant* because it determines the number of real solutions to $ax^2 + bx + c = 0$.

- $b^2 - 4ac > 0$: there are 2 distinct real roots to $ax^2 + bx + c = 0$. In fact, the graph of $y = ax^2 + bx + c$ cuts through the x-axis at two distinct points.

- $b^2 - 4ac = 0$: there is 1 real root (2 repeated real roots) to $ax^2 + bx + c = 0$. The graph of $y = ax^2 + bx + c$ is tangent to the x-axis.

- $b^2 - 4ac < 0$: there is no real root to $ax^2 + bx + c = 0$. The graph of $y = ax^2 + bx + c$ does not cross nor touch the x-axis.

16 Determine the number of real roots to each quadratic equation.

(a) $x^2 + x + 1 = 0$ 　　　　(b) $x^2 + x - 1 = 0$ 　　　　(c) $3x^2 - x + 3 = 0$

When there is a phrase about *tangency*, we can easily connect the idea of discriminant to the equation. If the relevant discriminant is equal to 0, then there is only one solution to the equation, meaning that the parabola meets the given line at one point.

17 Find the values of $k \in \mathbb{R}$ when the graph of $y = x^2 + kx + 4$ is *tangent* to the x-axis.

4.6 Complex Numbers

What if the discriminant of the quadratic equation is less than 0? Then, in the previous section, we learned that the equation has no real root. On the other hand, it does have *complex* roots. The set of complex numbers, denoted by $\mathbb{C}$, has all of its numbers in the form of $a + bi$ where $a, b \in \mathbb{R}$. In fact, if $b = 0$, then we are left with only real numbers. This tells us that $\mathbb{R} \subset \mathbb{C}$. What is so peculiar about complex numbers? Well, they consist of $i = \sqrt{-1}$. [1] Let's look at how algebraic simplification works for complex numbers.

Example

Simplify $\sqrt{-144}$ by using the imaginary number i.

> **Solution**
> $\sqrt{-144} = \sqrt{-1 \cdot 12 \cdot 12} = \sqrt{-1} \cdot \sqrt{12} \cdot \sqrt{12} = 12i$

Complex number arithmetic is not that difficult. We simply add real parts to real ones and imaginary parts to imaginary ones. What are the real parts and imaginary parts? Given a complex number $z = a + bi$, then the real part of z, denoted by $Re(z)$, is a. Similarly, the imaginary part of z, denoted by $Im(z)$, is b.

Example

Simplify $(3 + 4i) + (-2 - 3i)$.

> **Solution**
> $(3 + 4i) + (-2 - 3i) = (3 - 2) + (4 - 3)i = 1 + i.$

---

18 Simplify the following expressions.

(a) $(-3 + 5i) + (2 - i)$

(d) $(-1 + 3i)^2$

(b) $(1 + i)(4 - 2i)$

(e) $(1 + \sqrt{-3}) - \sqrt{-27}$

(c) $(3 - i)^2$

(f) $(2i - 1)3i$

---

[1] Normally, we do not allow any negative number inside the radical with even index. But, in the world of complex numbers, everything is possible.

The Fundamental Theorem of Algebra states that there are at most n complex roots to

$$a_n x^n + a_{n-1} x^{n-1} + \cdots + a_1 x + a_0 = 0$$

, where $a_0, a_1, \cdots, a_n$ are real numbers. By this statement, we draw two important conclusions.

- If all of the coefficients are *real* numbers and $f(a+bi) = 0$, then $f(a-bi) = 0$.

- If all of the coefficients are *rational* and $f(a+\sqrt{b}) = 0$, then $f(a-\sqrt{b}) = 0$.

We say that $a+bi$ and $a-bi$ are *complex conjugates*. In other words, they are the solutions to same polynomial equations. Same goes form $a+\sqrt{b}$ and $a-\sqrt{b}$ are *radical conjugates*. In short, if coefficients are real(or rational), then the conjugates are the solutions to the same polynomial equations (in this topic, quadratic equations).

19 Rationalize the denominators.

(a) $\dfrac{5+2i}{3i}$

(b) $\dfrac{2i}{2-i}$

(c) $\dfrac{1-i}{2+i}$

20 Find the factors of each expression, including complex coefficients. For notational convenience, we usually write z instead of x for complex factorization. However, in this question, we would stick with x.

(a) $x^2 + 4$

(b) $4x^2 + 9$

21 Find *all* solutions to each quadratic equation.

(a) $x^2 + 4x + 5 = 0$

(e) $2x^2 - 2x + 6 = 0$

(b) $-x^2 + 4x - 10 = 0$

(f) $3x^2 - 2x + 5 = 0$

(c) $2x^2 - 5x + 3 = 0$

(g) $4x^2 + 49 = 0$

(d) $-4x^2 - 6x + 1 = 0$

(h) $x^2 + x + 1 = 0$

4.7 Application of Quadratic Function and Equations

The first application of quadratic function is to graph the quadratic function with absolute value. What happens if we put absolute value on the quadratic function?

- The portion of $y = f(x)$ *below* the x-axis is reflected about the x-axis.

- The portion of $y = f(x)$ *above* the x-axis is unchanged.

The idea of sketching the graph of $y = |f(x)|$ begins with graphing the original graph of $y = f(x)$. Then, reflect the required parts about the x-axis, if needed.

22 Sketch the graph of $f(x) = |x^2 - x - 2|$.

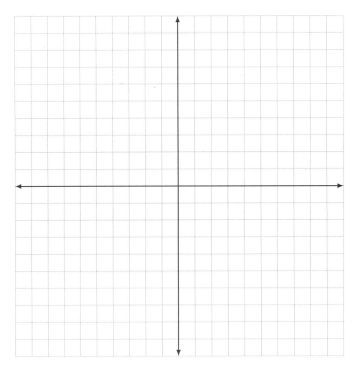

(a) Locate the x-intercepts.

(b) Locate the x-coordinate of the relative maximum[2].

---

[2]Relative maximum is the relatively maximum point around its neighborhood.

The second application of quadratic equation is to find the range with the restricted domain. Be careful about restricted domain. Let's assume that the vertex of $y = ax^2 + bx + c$ is located at (h, k). Focus on the x-coordinate of the vertex. Imagine restricting the domain $m \le x \le n$.

- If $m < n < h$, then $f(m)$ is maximum and $f(n)$ is minimum, assuming $a > 0$. On the other hand, $f(m)$ is minimum and $f(n)$ is maximum, assuming $a < 0$.

- If $m < h < n$, then $f(h)$ is minimum and the larger of $f(m)$ and $f(n)$ is maximum, assuming $a > 0$. On the other hand, $f(h)$ is maximum and the smaller of $f(m)$ and $f(n)$ is minimum, assuming $a < 0$.

- If $h < m < n$, then $f(m)$ is minimum and $f(n)$ is maximum, assuming $a > 0$. On the other hand, $f(m)$ is maximum and $f(n)$ is minimum, assuming $a < 0$.

$\boxed{23}$ Find the range of $x^2 - 2x - 2$ for the domain $-2 \le x \le 4$. Sketch the graph in the following xy-plane.

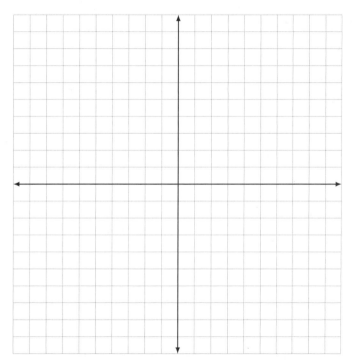

(a) Find the y-value for the endpoints.

(b) Find the y-value of the vertex.

24 Sketch the graph of the function $f(x) = |x(x-2)|$ for the domain $-1 \le x \le 3$.

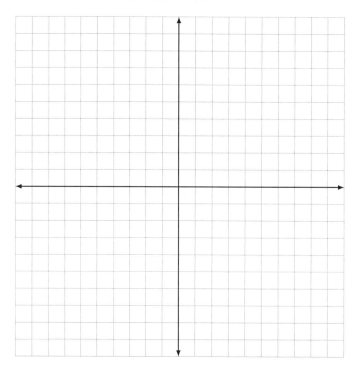

(a) Find the range of $f(x)$.

(b) Find the possible values of k for $f(x) = k$ has more than two solutions.

The application of quadratic equations cannot dispense with word problems. Note that the word problem involves real life application, so the variable normally takes a positive value.

25 A rectangular fence is made against a straight wall, using three lengths of the fencing, two of length x meters. The total length of the fencing available is 50 meters.

(a) Show that the area enclosed is given by $A(x) = 50x - 2x^2$.

(b) Find the maximum possible area which can be enclosed and the value of x for the area.

4.8 Quadratic Inequalities

Quadratic inequality is a bit trickier than quadratic equation. There are two methods of solving the inequality. The first method is beloved by many teachers at school, known as *sign analysis*. This is great for practicing caseworks. The second method is quicker and bit more visual. It requires graphing the function $y = ax^2 + bx + c$ and comparing it with the given horizontal line, normally the x-axis. For instance, $2x^2 + 3x + 1 < 0$ calls for comparing the graph of $y = 2x^2 + 3x + 1$ and $y = 0$ (the x-axis) and locate the portion of the graph of $y = 2x^2 + 3x + 1$ lower than the x-axis.

Example

Solve $x^2 + x < 0$.

Solution

1. Since $x(x+1) < 0$, we could graph it in the coordinate plane.

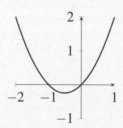

The portion of the graph lower than the x-axis corresponds to $(-1, 0)$.

#2. Since $x^2 + x = x(x+1) < 0$, locate the important points $x = -1$ and $x = 0$.

Case 1. $x < -1 : (-)(-) > 0$, so this is not part of the solution set.
Case 2. $-1 < x < 0 : (-)(+) < 0$, which is included in the solution set.
Case 3. $0 < x : (+)(+) > 0$, which is excluded from the solution set.

Combining all three caseworks, we get the solution set $-1 < x < 0$.

What if the expression cannot be factorized? We may complete the square to locate the graph in the xy-plane or use discriminant method. Have a look at the following example.

26 Show that $3x^2 - 2x + 3 > 0$, using the discriminant method. (Hint: Show that the graph never meets the x-axis.)

27 For what set of x-values is $3x^2 - 4x \leq -1$?

The existence of *real roots* asks us to use the discriminant naively when solving a quadratic equation problem. If there is at least one real root, it indicates that the discriminant is non-negative. Have a look at the next example.

28 For what range of values k will the equation $x^2 - kx + 2x + k^2 + 3k - 3 = 0$ have real roots?

Graphing helps us locate where the solutions might be, but it may not give the *exact* location. In such case, the best way is to use *algebraic* method to find the specific values.

29 Find the domain of x for which $|x^2 - 3x - 7| \geq 3$. The graph of $y = |x^2 - 3x - 7|$ and $y = 3$ are given by the following figures.

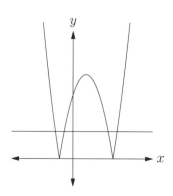

1 Given that $4x^2 - 16x + 15 = a(x-p)^2 + q$ for all values of x, find the values of a, p and q. Hence, state the coordinates of the minimum point of the curve $y = 4x^2 - 16x + 15$.

2 Find the equation of the quadratic function whose vertex is $(-2, 3)$ that passes through $(-1, 5)$.

3 Given that $y = kx^2 - 4x + 3k$, express y in the form $a(x - p)^2 + q$, where a, p and q are in terms of k. Hence, find the value of k if the maximum value of y is 4.

4 Find the values of p for which the x-axis is a tangent to the curve $y = x^2 + px - p + 3$. For each of these values, find the coordinates of the point of tangency.

5 What is the least value k can take if the roots of the equation $x^2 - 2kx + k^2 = 3 + x$ are real?

6 If the line $y + 2x = p$ does not intersect the curve $y^2 = x + p$, find the range of values of p.

7 Find the solution set of each of the following quadratic inequalities.

(a) $x(x-2) < 3$

(b) $x^2 > 4x + 12$

(c) $(1-x)^2 \geq 17 - 2x$

8 Find the set of values of m for the equation $2x^2 + 4x + 2 + m(x+2) = 0$ to have real roots.

9 The quadratic function $y = (k-6)x^2 - 8x + k$ cuts the x-axis at two points and has a minimum point. Find the range of values of k.

10 Given that $y = 2x^2 + px + 16$ and that $y < 0$ only when $2 < x < k$, find the value of p and k.

11 The equation $m^2x^2 - (m+n)x + n = 0$ has two equal real roots. Show that $n^2 + 2(m - 2m^2)n + m^2 = 0$. If n is real, find the range of values of m.

12 Find the range of values of k if $kx^2 + 8x > 6 - k$ for all real values of x. (Assume that k is positive.)

1 $a = 4$, $p = 2$, and $q = -1$. Hence, the minimum point is $(2, -1)$.

2 $y = 2x^2 + 8x + 11$.

3 $a = k$, $p = \dfrac{2}{k}$, $q = \dfrac{3k^2 - 4}{k}$. The value of k is $-\dfrac{2}{3}$.

4 $p = -6, 2$. At $p = -6$, $(3, 0)$ is the point of tangency. At $p = 2$, $(-1, 0)$ is the point of tangency.

5 $k = -\dfrac{13}{4}$.

6 $p < -\dfrac{1}{24}$.

7

(a) $\{x \in \mathbb{R} : -1 < x < 3\}$ (b) $\{x \in \mathbb{R} : x < -2 \text{ or } 6 < x\}$ (c) $\{x \in \mathbb{R} : x \leq -4 \text{ or } 4 \leq x\}$

8 $(-\infty, 0] \cup [8, \infty)$.

9 $6 < k < 8$.

10 $k = 4$ and $p = -12$.

11 Set discriminant equal to 0. Hence, the range of values of m is $(-\infty, 0] \cup [1, \infty)$.

12 $8 < k$.

유하림 저자에게 듣는 공통 Q&A

Q. 이 교재는 어느 정도 수준의 학생들에게 적합한가요?

A. Algebra 2를 처음 공부하려고 하는 학생들에게 적합하며, 난이도는 기본으로 설정되어 있습니다. 이러한 이유로 여러 학생들에게 모두 전달이 가능합니다. 상위권 학생들을 위한 교재라기 보다는, Algebra 2를 처음 공부하는 학생들에게 적합합니다. 단, 선생님이 개인지도하거나 다수의 학생에게 수업을 진행할 때, 개념을 확장 시키기에 적합한 기본 문형들이 대거 포함되어 있습니다.

Q. Algebra 2를 어느 시점에 학습하는 것이 좋을까요?

A. 학생들마다 다르겠지만, 국내 학생들 기준으로 중등부 과정의 개념 학습이 어느 정도 끝난 학생이면 바로 시작하는 것을 추천합니다. 해외 기준으로 보면, Algebra 1과 Geometry가 끝난 학생들이 시작하는 것이 좋으나, 병행하고자 하는 학생에게 다음과 같이 추천합니다.

이 교재를 활용할 시, Algebra 1의 경우, Quadratics를 시작하는 시점에서 Algebra 2를 시작하는 것도 나쁘지 않은 선택이며, Geometry와 동시에 진행하는 것도 저학년 학생들(7, 8학년)에게 큰 무리가 없습니다.

Q. The Essential Guide to Algebra 2는 타교재 대비 어떤 부분이 다른가요?

A. 우선, 유사한 부분부터 말씀드리면, 타 교재들과 비교해보았을 때 다루고 있는 개념의 폭 부분에서는 큰 차이를 보이지 않습니다. 그러나, 이 교재는 30시간 기준으로 현장 강의에서 다루기에 적합하도록 집필한 교재로, 불필요한 내용들은 최대한 뺀 교재입니다. Algebra 2 개념 내용을 빠르고, 효율적으로 전달하기 위해 집필한 교재일 뿐 아니라, 학교 내신 시험에 꼭 필요한 내용들 위주로 문제와 그 풀이를 집필하였습니다. 특히, 학교 내신 시험 및 퀴즈에서 학생들이 풀이를 적을 때 빠뜨리면 감점요인이 되는 부분들을 강조하기 위해서 사용하는 예시들이 많기 때문에, 이 교재를 주도적으로 학습하는 학생들 입장에서는, 풀이를 꼼꼼히 적는 훈련을 하며 교재를 활용한다면, 저자가 의도한 효과를 십분 활용할 수 있습니다.

Topic 5

Polynomials and Polynomial Function

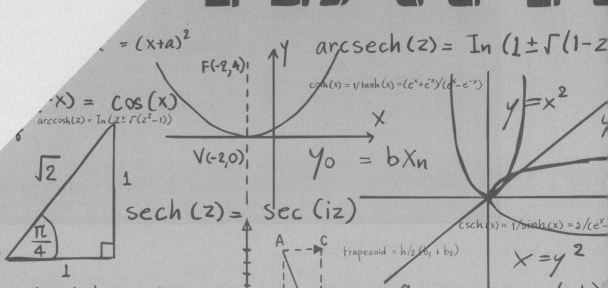

5.1 Classification of Polynomials

Given a polynomial of x,

$$p(x) = a_n x^n + a_{n-1} x^{n-1} + \cdots + a_1 x + a_0$$

1. n is the degree.

2. a_n is the leading coefficient.

Also, we can classify a polynomial by the number of terms.

1. Three-term polynomial is called trinomial.

2. Two-term polynomial is called binomial.

3. One-term polynomial is called monomial.

One more vocabulary for polynomial is the name of important polynomials.

- A polynomial of degree 1 is known as linear.

- A polynomial of degree 2 is known as quadratic.

- A polynomial of degree 3 is known as cubic.

- A polynomial of degree 4 is known as quartic.

- A polynomial of degree 5 is known as quintic.

Example

Classify the following polynomial by degree and the number of terms.

$$5x^2 - 4 + 6x^5$$

Solution
The degree is 5, and it is trinomial. When we write a polynomial, it is written in decreasing order. Hence, the proper form must be $6x^5 + 5x^2 - 4$.

1 Rewrite the following polynomial in standard form

$$x(x+5) - 5(x+5)$$

and solve the following questions.

(a) Classify it by degree.   (b) Classify it by the number of terms.

2 Write the following polynomials in standard form. Classify each by its degree and its number of terms.

(a) $x^3(x^2-x+1)$

(b) $5-3x^3+4+x^3$

3 Which of the following is a polynomial function? (Hint : There is an answer to this question. Think deeply about the polynomial form.)

(A) $f(x)=\ln(x^2+2)$
(B) $f(x)=|x^2-x+3|$
(C) $f(x)=\sin(x)$
(D) $f(x)=e^{2x}+e^x+1$

5.2   End-Behavior of Polynomial Function

The end-behavior depends on degree and leading coefficient.

1. If the degree is odd, then the end-behavior is up-down or down-up.

2. If the degree is even, the end-behavior is down-down or up-up.

Example

Determine the end behavior of $y = -2x^3 + 5x^2$.

> **Solution**
> Since the degree of the polynomial function is 3, the end-behavior goes opposite direction. As x goes to $-\infty$, y goes to ∞. On the other hand, as x goes to ∞, then y goes to $-\infty$.

4 Determine the end-behavior of $y = 3x^4 + 6x^3 - x^2 - 12$.

(a) The leading coefficient is (positive/negative).

(b) The degree is (even/odd).

(c) Hence, the end-behavior is (up-up / up-down / down-up / down-down).

5 Determine the end-behavior of $y = 20 - 5x^6 + 3x - 11x^3$.

(a) The leading coefficient is (positive/negative).

(b) The degree is (even/odd).

(c) Hence, the end-behavior is (up-up / up-down / down-up / down-down).

5.3　The Shape of Cubic Graph

Parent function of a cubic function is $y = x^3$, whose graph is symmetric about the origin[1], which is known to be the inflection point. In cubic graphs, the maximum number of turning points is 1 less than the degree of the given polynomial.

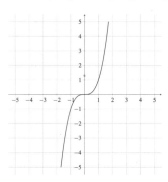 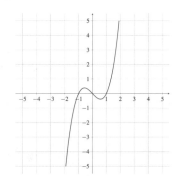 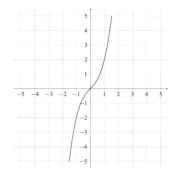

6 Describe the shape of the graphs of the following cubic function by the end-behavior and the number of turning points.

(a) $y = x^3 + 4x$

(b) $y = -2x^3 + 3x - 1$

(c) $y = 5x^3 + 6x^2$

7 Sketch the graph of $y = (x-1)(x-2)(x-3)$.

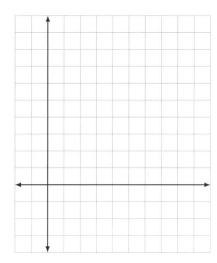

[1]Point symmetry equals reflection about $180°$.

5.4 Table Values and Polynomial Degrees

The following example will illustrate how to find the polynomial degree with the given data.

Example

x	-2	-1	0	1	2
y	-16	1	4	5	16

Compute the first consecutive difference $17, 3, 1, 11$. Then, the consecutive difference of the difference is $-14, -2, 10$. Notice the difference of the second difference. It's $12, 12, \ldots$. Hence, it must be cubic function because we have the third difference constant.

8 Determine the degree of the polynomial function with the given data.

x	-3	-2	-1	0	1	2	3
y	-27	-8	-1	0	1	8	27

9 Determine the degree of the polynomial function with the given data.

x	-4	-3	-2	-1	0	1	2	3	4
y	249	76	13	0	1	4	21	88	265

5.5 Finding x-intercepts

We can write a polynomial in factored form so that we can locate the x-intercepts. Just as the intercept form of quadratic equations helps us understand its graph, the intercept form of polynomials helps us visualize how the graph would look like.

> **Example**
>
> $$\begin{aligned} x^3 + 3x^2 + 3x + 1 &= (x^3 + 2x^2 + x) + (x^2 + 2x + 1) \\ &= x(x^2 + 2x + 1) + (x^2 + 2x + 1) \\ &= (x+1)(x+1)^2 \\ &= (x+1)^3 \end{aligned}$$
>
> The graph of $y = (x+1)^3$ cuts through $x = -1$ once, and there is no other x-intercept.

10 Factor the following polynomials.

(a) $x^3 - 2x^2 + x$

(b) $x^4 - 4x^2 + 4$

11 Write the following polynomials in factored form.

(a) $x^3 + 7x^2 + 15x + 9$

(b) $-3x^3 + 18x^2 - 27x$

5.6 Finding Zeros

Given $y = (x+1)(x-1)(x+3)$, we can easily find the x-intercepts, i.e., $x = -1, 1, -3$. Based upon the x-intercepts and the end-behavior, we can sketch the graph.

12 Sketch the graph of $y = x(x-2)(x+5)$.

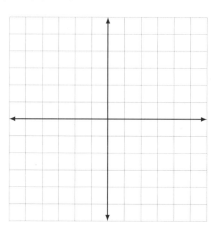

13 Find the zeros of the following functions and sketch the graph of the function.

(a) $y = (x+4)^2(x+1)$

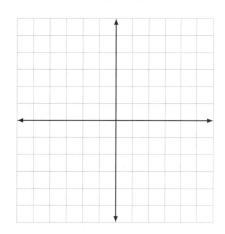

(b) $y = (x+4)^2(x+1)^2$

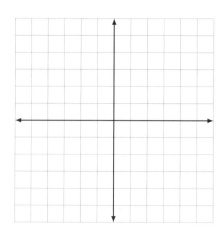

5.7 Given Zeros

Given zeros, we can always write a polynomial function. For instance, if $x = -1, 3, 4$, then a polynomial function $(x+1)(x-3)(x-4)$ may be used. Such technique of using polynomials when some number of solutions are provided is called "polynomial interpolation."

14 Write a polynomial function in standard form with the given zeros.

(a) $x = -3, 0, 0, 5$ (b) $x = 4, 2, -3, 0$ (c) $x = 1, 1, 2, 2$

Given $y = (x-1)^2(x-3)^4$, there are two zeros, i.e., $x = 1, 3$. How do we count the number of times that a root appears in a given polynomial function? We use the concept of multiplicity. For example, given $y = (x-3)^2$, we say that $x = 3$ has the multiplicity of 2.

15 State the multiplicity of multiple zeros of the expression $y = 9x^3 - 81x$.

1 Write each polynomial in standard form, and classify it by degree and the number of terms.

(a) $\dfrac{3x^2 + 4x + 1}{5}$

(b) $\dfrac{2 + 3x^5}{5}$

2 Determine the end-behavior of the graph of each polynomial function.

(a) $y = 4 + 4x^6 + 2x - 11$

(b) $y = 11 - 2x + 3x^2 - 4x^6$

$\boxed{3}$ What is the number of terms in the polynomial $(3x - 3)(4x^2 - 1)$?

(A) 2
(B) 3
(C) 4
(D) 5
(E) 6

$\boxed{4}$ Factor the following polynomials.

(a) $x^3 - 7x^2 + 15x - 9$ (b) $x^3 - 3x^2 + 3x - 1$

$\boxed{5}$ Find the zeros and multiplicity of each zero for the polynomial function $f(x) = x^4 - 8x^2 + 16$.

6 Sketch the graph of the following cubic functions.

(a) $y = (x-1)^2(x-2)$

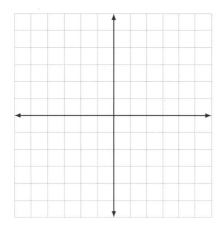

(b) $y = 2x^3 + 2x$

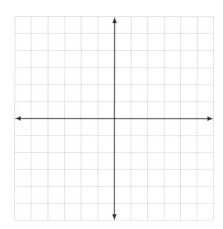

(c) $y = x(x-1)(x+2)$

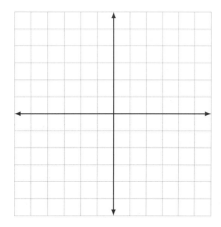

(d) $y = -x^3$

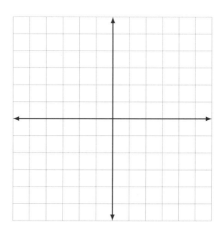

1

(a) $\frac{3}{5}x^2 + \frac{4}{5}x + \frac{1}{5}$ has the degree of 2, and there are three terms, known as a trinomial.

(b) $\frac{3}{5}x^5 + \frac{2}{5}$ has the degree of 5, and there are two terms, known as a binomial.

2

(a) up-up   (b) down-down

3 (C)

4

(a) $(x-1)(x-3)^2$ (b) $(x-1)^3$

5 Zeros are 2 and -2. The multiplicity of each zero is 2.

6

(a)   (b)   (c)   (d)

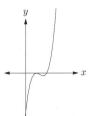 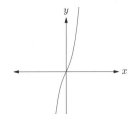 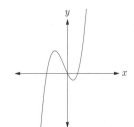 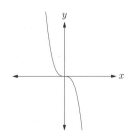

Algebra 2와 수학 경시대회 관련한 Q&A

Q. AMC 10/12 또는 9, 10학년이 출전할 수 있는 경시대회 준비를 위해, Algebra 2는 반드시 학습해야 하나요?

A. AMC 10/12 공통 주제로 Algebra 2에서 다루는 개념들을 활용한 문제들은 자주 출제되는 편입니다. 예를 들어, rational function을 활용한 AM-GM inequality는 자주 출제된다고는 할 수 없지만, 출제 빈도수가 적은 편도 아니므로 반드시 알아야 하는 내용입니다. 특히, counting and probability 부분은 AMC 10/12뿐 아니라, AMC 8에서부터 꾸준히 출제되고 있는 개념 부문이므로, 개념을 한번 잡고 가는 것이 필수적입니다.

Q. 이 교재로 수학 경시대회 준비가 가능할까요?

A. 이 교재는 Algebra 2를 처음 배우고자 하는 학생들을 위해, 개념 입문편이라고 볼 수 있습니다. 개념을 한 번도 학습해보지 않은 학생들에게는 추천하는 교재입니다. 즉, Algebra 2를 한 번도 수강해보지 않은 학생들이라면, 이 교재를 통해 우선 학습하는 것을 추천합니다. 다만, 수학 경시대회 대비를 위한 교재로 활용하기에는 응용의 폭 및 깊이 부분에서 부족하니, 저자의 Essential Math Series 중에서 The Essential Guide to Competition Math : Counting and Probability를 선택하여 학습하는 것이 오히려 더 효과적일 수 있습니다.

Q. 한국 수학 커리큘럼상 수1, 2 개념을 학습한 상태면, Algebra 2 개념 학습 없이도, 수학 경시대회 준비가 바로 가능할까요?

A. 이 부분은 학생마다 다르게 적용합니다. 예를 들어, 수1, 2 개념을 학습하여서, Algebra 2에 있는 개념들이 대부분 학습된 상태일 때, 수학 경시대회 준비는 가능하겠으나, Algebra 2 개념을 학습할 때 배우는 특정한 방식들에 대해서 노출되지 않았다면, 개념에 공백이 있을 수 있습니다. 이러한 이유로, 학생들 개별 상황에 맞추어서 대비하는 것이 좋습니다.

Topic 6

Application of Polynomials

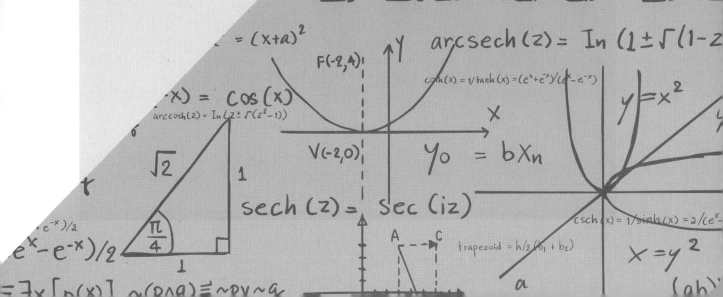

6.1 Factorization

The sum of cubic difference or sum oftentimes helps students to manipulate a given algebraic expression. The following example shows the expansion of well-known cubic expressions.

Example

1. $(a+b)^3 = a^3 + 3a^2b + 3ab^2 + b^3$

2. $a^3 + b^3 = (a+b)^3 - 3ab(a+b) = (a+b)(a^2 - ab + b^2)$

3. $(a-b)^3 = a^3 - 3a^2b + 3ab^2 - b^3$

4. $a^3 - b^3 = (a-b)^3 + 3ab(a-b) = (a-b)(a^2 + ab + b^2)$

$\boxed{1}$ Find the real or imaginary solutions of each polynomial equation by factorization.

(a) $8x^3 - 27 = 0$

(b) $x^4 - 11x^2 + 10 = 0$

$\boxed{2}$ Solve for x for the following polynomial equations.

(a) $x^4 + 3x^2 - 4 = 0$

(b) $2x^3 = 250$

3 Solve for x for the following polynomial equations.

(a) $3x^4 = 3x$

(e) $x^4 + 6 = 7x^2$

(b) $x^4 - 13x^2 - 14 = 0$

(f) $x^4 - x^2 - 132 = 0$

(c) $x^4 - 16 = 0$

(g) $x^4 - 25 = 0$

(d) $x^5 - x^3 - 12x = 0$

(h) $3x^4 + 36x^2 = 24x^3$

6.2 Long Division

We could perform long division to polynomials, just as we do to numbers. The purpose of long division begins with curiosity for finding factors. However, the in-depth application of long division tells us the possible location of x-intercepts. Let's look at the following example.

Example

Explain why there is no x-intercept greater than $x = 1$ for $f(x) = x^3 + 1$.

Solution
If you perform a long division on $x^3 + 1$ by a linear divisor of $x - 1$, we get $x^3 + 1 = (x-1)(x^2 + x + 1) + 2$. If $x > 1$, then $x - 1 > 0$ and $x^2 + x + 1 > 0$. Hence, there is no way that $x^3 + 1$ will ever get closer to 0 if $x > 1$. This means that we have an upperbound of x-intercept by performing one long division process.

$\boxed{4}$ Perform long divisions.

(a) Divide $x^2 - 5x - 12$ by $x + 3$.

(b) Divide $x^3 + 4x^2 - 2x - 1$ by $x - 1$.

(c) Divide $x^3 - 5x^2 + 3x + 2$ by $x + 2$.

$\boxed{5}$ Find the quotient when $x^3 + x^2 + x + 1$ is divided by $x + 1$.

6 Divide the following polynomials.

(a) $x^7 - 1$ by $x - 1$

(b) $x^4 + 3x^2 - 4x + 1$ by $2x - 1$

(c) $x^3 + 11x^2 - 13x + 24$ by $x + 1$

7 Find the remainder when $4x^3 + 4x^2 + x - 1$ is divided by $x + 2$.

6.3 Factor Theorem

Other than long division, factor theorem tells us whether the given binomial is a factor of the larger polynomial. Let's say we divide $p(x)$ by $x - a$. Then, there must be a quotient, $Q(x)$, and a remainder r, in the following form.

$$p(x) = (x - a)Q(x) + r$$

What do we know from this expression? The first impression that we get is that we automatically know what the remainder is. In fact, the remainder r is simply equal to $p(a)$ because

$$\begin{aligned} p(a) &= (a - a)Q(a) + r \\ &= 0 \cdot (Q(a)) + r \\ &= 0 + r \\ &= r \end{aligned}$$

Notice that if $r = 0$, then we know $p(a) = 0$, which implies that $x - a$ is a factor of $p(x)$.

8 Determine whether the following linear term is a factor of $x^3 + 3x^2 - 10x - 24$.

(a) $x - 3$ (b) $x + 4$ (c) $x + 2$

9 How can you be certain that $x^3 + 3x^2 + 4x + 1$ has no positive factor? Show this using factor theorem.

6.4 Synthetic Division

One of the most interesting procedures we learn in polynomial application is synthetic division. Synthetic division, easier than long division, can only be applied when the divisor is linear. Here, unlike many other regular textbooks for Algebra 2, we would like to see *why* we learn synthetic division. Is it merely a simpler process than long division? If that is the case, then it seems that our learning is quite mechanical. However, this is not the only reason why we learn synthetic division.

Synthetic divison literally translates polynomial left and right. This saves us lots of energy and time when we solve some challenging polynomial problems in math competitions. Have a look at the following example.

Example

If $p(x) = x^3 - 1 = (x-1)(x^2+x+1)$, determine the coefficients of $q(x)$ such that $q(x+1) = p(x)$.

Solution

$$x^3 - 1 = (x+1)(x^2 - x + 1) - 2$$
$$= (x+1)((x+1)(x-2) + 3) - 2$$
$$= (x+1)((x+1)((x+1) - 3) + 3) - 2$$
$$= (x+1)^3 - 3(x+1)^2 + 3(x+1) - 2$$

This implies that $q(x) = x^3 - 3x^2 + 3x - 2$, which is equal to $(x-1)^3 - 1$. What just happened? If you perform synthetic division multiple times with -1 to $1, 0, 0, -1$, you get the number $-2, 3, -3$, and 1 as the last number in each step. They all correspond to the coefficients of the transformed polynomial, which is simply 1 unit translated to the right.

---

10 Divide the following cubic polynomials by the given linear factor using synthetic division.

(a) $x^3 - 8x^2 + 17x - 10$ by $x - 5$.

(b) $x^3 - 5x^2 - 7x + 25$ by $x - 5$.

(c) $x^3 + 2x^2 + 5x + 12$ by $x + 3$.

11 Compare the quotients produced by long division and synthetic division when $x^3 - 4x^2 + 3x + 1$ is divided by $2x - 1$.

Look at the following example to find the connection between the remainder theorem and the synthetic division.

12 If $p(x) = 3x^3 - 4x^2 - 5x + 1$ is divided by $x - 2$, what is the remainder?

The Fundamental Theorem of Algebra states that every nonconstant polynomial has at least one root. Thus, there is at least one value a such that $f(a) = 0$. This a may be real, imaginary, rational, or irrational, but the Fundamental Theorem of Algebra assures us that at least one such root exists. Unfortunately the proof is a bit too complex for this book, but we shall put the theorem to good use by showing that any degree n polynomial has exactly n roots. This means we can write any polynomial $f(x)$ as

$$f(x) = a_n x^n + a_{n-1} x^{n-1} + \cdots + a_1 x + a_0$$
$$= a_n (x - r_1)(x - r_2) \cdots (x - r_n).$$

The r_is are the roots of the polynomial and they are not necessarily real or rational. It should be clear why $f(r_i) = 0$.

13 A polynomial function $P(x)$ with *integer coefficients* has the given roots. Find the two additional roots to $P(x) = 0$.

(a) $1 - i$ and $\sqrt{5}$ (b) $2 - \sqrt{3}$ and $-3 + \sqrt{7}$

6.5   Rational Root Theorem

For the rational roots of a polynomial, there is a method we can use to narrow the search. Although there are infinitely many rational numbers we could guess as roots of $f(x) = 0$, the only ones which have a chance of being roots are given by the Rational Root Theorem. For any polynomial

$$f(x) = a_n x^n + a_{n-1} x^{n-1} + \cdots + a_0$$

with integer coefficients, all rational roots are of the form p/q, where $|p|$ and $|q|$ are relatively prime integers, p divides a_0 evenly, and q divides a_n evenly.

Oftentimes, problem-solvers may use the rational root theorem with the "location principle," which states that if successive integers are plugged inside the polynomial, there must be a root between two integers where the sign of polynomial expression changes.

14 Use the rational root theorem to list all possible rational roots for the following equations. Hence, find any actual rational root.

(a) $x^3 - 6x^2 + 11x - 6 = 0$

(b) $2x^3 - 5x^2 + 4x - 1 = 0$

15 Factorize the following polynomial equations using rational root theorem.

(a) $x^3 - 4x^2 + 5x + 10 = 0$

(b) $x^3 + x - 10 = 0$

16 Use the rational root theorem to solve the following cubic polynomial equations.

(a) $x^3 - 8x^2 + 9 = 0$

(d) $x^3 + x^2 - 27x + 25 = 0$

(b) $x^3 + x^2 - 5x + 3 = 0$

(e) $x^3 - 18x + 27 = 0$

(c) $x^3 + x^2 - 34x + 56 = 0$

(f) $x^3 - 5x^2 + x - 5 = 0$

6.6 Two guidelines for finding roots

The first is Descartes' Rule of Signs, which gives us a method to count how many positive and how many negative roots there are. We look at the number of alternating signs of coefficients. In short, the number of sign changes in the coefficients of $f(x)$ (meaning we list the coefficients from first to last and count how many times they change from positive to negative) tells us possibly maximum number of positive roots, whereas and the number of sign changes in the coefficients of $f(-x)$ gives us possibly maximum number of negative roots. Hence, for $f(x) = 3x^5 + 2x^4 - 3x^2 + 2x - 1$, there are at most 3 positive roots and at most 2 negative roots, since $f(-x) = -3x^5 + 2x^4 - 3x^2 - 2x - 1$).

The second is finding upper and lower bounds. Suppose we use synthetic division to find $f(x)/(x-c)$ where $f(x)$ has a positive leading coefficient and $c \geq 0$ as below:

$$
\begin{array}{r|rrrr}
3 & 1 & -1 & 2 & 6 \\
  &   & 3 & 6 & 24 \\
\hline
  & 1 & 2 & 8 & 30 \\
\end{array}
$$

If all the resulting coefficients in the quotient are positive (including the remainder), as in the example above, then no roots are greater than c. (Why?) This c is called an upper bound on the solutions since no roots can be higher. Similarly, if $c < 0$ and the coefficients of the quotient and remainder alternate in sign, then there is no root smaller than c (which we then call a lower bound for the roots). Locating upper and lower bounds will often help you shorten your search for roots.

17 Use Descartes' Rule of Signs to tell the number of positive real roots and negative real roots for each polynomial functions.

(a) $p(x) = 4x^3 + x^2 - x - 13$

(b) $p(x) = x^3 + x^2 + 2x - 7$

18 In order to find one of the roots of

$$
f(x) = 2x^4 - 15x^3 + 15x^2 + 20x - 12.
$$

I start with $x = 1$. After finding $f(1) = 10$, what should I try next? (Hint: Think about the location principle stated in section 6.5.)

1 Use the rational root theorem to find all the real zeros of the following polynomial functions.

(a) $x^3 + 2x^2 - 5x - 6 = 0$

(d) $x^3 - 2x^2 + 2x - 1 = 0$

(b) $3x^3 + 4x^2 - 7x + 2 = 0$

(e) $x^3 - 2x^2 - 5x + 10 = 0$

(c) $x^4 + x^3 - 3x^2 - x + 2 = 0$

(f) $x^3 + 1 = 0$

2 Use the remainder theorem to find the remainder when $f(x)$ is divided by the following linear term.

(a) $x^2 - x^3 + 2x - 1$ by $x - \dfrac{1}{2}$

(b) $x^4 + 10x^2 - 3x - 1$ by $x + 1$

3 Completely factorize $3x^4 + 5x^3 + 25x^2 + 45x - 18$.

$\boxed{4}$ Use the given complex zeros to find the other zeros of the following polynomial functions.

(a) $p(x) = x^5 + 3x^4 + 25x + 75$ where $x = \sqrt{5}i$ is a zero.

(b) $p(x) = x^4 + 13x^2 + 36$ where $x = 2i$ is a zero.

$\boxed{5}$ Find the first-hand bounds of the real zeros of each polynomial function. (You may use rational root theorem or intermediate value theorem[1] along with location principle.)

(a) $f(x) = x^3 - x^2 - 5x + 2$ (b) $f(x) = x^3 + x^2 + 10x - 5$

---

[1] If $y = f(x)$ is *continuous* function and $f(a)f(b) < 0$ for $a < b$, there exists at least one x-intercept between a and b. This is a brief application of intermediate value theorem, also known as IVT. Any polynomial function is continuous function, so IVT works fine for polynomial functions.

6 List all possible rational zeros of $f(x) = 5x^5 + x^4 - x^3 + 2x - 1$.

7 Use Descartes' rule of signs to find out the number of positive or negative real zeros for the following polynomial functions.

(a) $p(x) = 4x^5 - x^2 + x - 1$ 　　　　　　　　　　 (b) $p(x) = x^3 + x^2 + x - 3$

1

(a) $x = -1, -3, 2$

(b) $x = \dfrac{2}{3}, -1 \pm \sqrt{2}$

(c) $x = 1(\text{double}), -2, -1$

(d) $x = 1, \dfrac{1 \pm \sqrt{3}i}{2}$

(e) $x = 2, \pm\sqrt{5}$

(f) $x = -1, \dfrac{1 \pm \sqrt{3}i}{2}$

2

(a) $\dfrac{1}{8}$ (b) 13

3 $(3x - 1)(x + 2)(x^2 + 9)$

4

(a) The zeros are $\pm\sqrt{5}i$, $\pm\sqrt{-5}i$, -3. (Later in Precalculus, we learn how to convert $\sqrt{5}i$ into $(1 + i)\sqrt{\frac{5}{2}}$.)

(b) The zeros are $\pm 3i$ and $\pm 2i$.

5

(a) At $x = 0$, $f(0) = 2 > 0$. Also, at $x = 1$, $f(1) = -3 < 0$. Hence, there must be a zero between $x = 0$ and $x = 1$.

(b) At $x = 0$, $f(0) = -5 < 0$. Similarly, $f(1) = 7 > 0$. Hence, there must be a zero between $x = 0$ and $x = 1$.

6 The possible candidates of x using rational root theorem are $x = \pm\dfrac{1}{5}, \pm 1$.

7

(a) There could be three positive real zeros or one positive real zero. On the other hand, there is no negative real zero.

(b) There is only one positive real zero. On the other hand, there could be two negative real zeros or no negative real zero.

Topic 7

Radical Expression and Radical Function

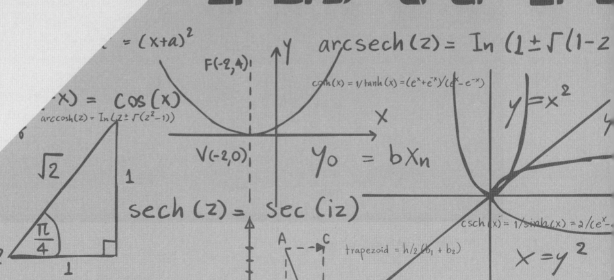

7.1 Introduction to Radicals

Radical expression is given by

$$x^{\frac{1}{n}} = \sqrt[n]{x}$$

where n is known as index. If n is even, then x should be non-negative. Otherwise, x is any real number.

Let's start with square roots.

Example

Find the square root of 196.

Solution
$$\sqrt{196} = \sqrt{2 \cdot 7 \cdot 2 \cdot 7} = \sqrt{2}\sqrt{2}\sqrt{7}\sqrt{7} = 2 \times 7 = 14.$$

This can be generalized to the nth root of x. For instance,

$$\sqrt[3]{8} = \sqrt[3]{2 \times 2 \times 2} = \sqrt[3]{2} \times \sqrt[3]{2} \times \sqrt[3]{2} = 2.$$

1 Find each root.

(a) $\sqrt{169}$

(b) $-\sqrt{36}$

(c) $\sqrt[3]{0.008}$

(d) $\sqrt[3]{-64}$

(e) $\sqrt{0.16}$

2 Find all real cube root(s) of the following numbers.

(a) 216

(b) -343

(c) 64

(d) $\dfrac{27}{1000}$

3 Find all the real fourth root(s) of the following numbers. If such root does not exist, write "does not exist."

(a) -81

(b) 256

(c) 64

(d) 81

There are certain types of simplification questions with respect to radical expressions. Utilize the following two bullet points for simplification.

- $\sqrt[n]{x^n} = |x|$ if n is even.

- $\sqrt[n]{x^n} = x$ if n is odd.

Example

Compute $\sqrt[3]{-8}$.

> **Solution**
> Since $-8 = (-2)^3$, we get $\sqrt[3]{(-2)^3} = -2$.

Example

Simplify the following expression $\sqrt{16x^2y^4}$.

> **Solution**
> $\sqrt{16x^2y^4} = \sqrt{16}\sqrt{x^2}\sqrt{y^4} = 4|x|y^2$.

$\boxed{4}$ Simplify the following radical expressions.

(a) $\sqrt[4]{\dfrac{x^4}{16}}$

(b) $\sqrt[6]{(x-y)^6}$

(c) $\sqrt{8(a+b)^4}$

(d) $\sqrt[4]{\dfrac{x^4}{64}}$

5 The volume V of a sphere with the radius of r is given by $V = \frac{4}{3}\pi r^3$.

(a) What is the radius of a sphere with volume 36π cubic inches? (If possible, justify why there is only one real value of r.)

(b) If the volume increases by a factor of 8, what is the new radius?

6 The velocity of a falling object can be found by $v^2 = 64h$, where v is the velocity (in feet per second) and h is the distance the object has already fallen. (Only for part (b), assume that the term *velocity* is used instead of *speed*. Normally, velocity can take negative values, while speed is its absolute value.)

(a) What is the velocity of the object after a 10-foot fall?

(b) How much does the velocity increase[1] if the object falls 20 feet rather than 10 feet?

---

[1]The correct expression must be either "velocity decrease" or "speed increase." For this question, solve it in terms of speed.

7.2 Multiplying Radical Expressions

When we multiply radical expressions, it is easy to break down radicands into fully factorized forms. Let's look at the following example.

> **Example**
>
> Multiply and simplify $\sqrt[3]{15}\sqrt[3]{75}$.
>
> > **Solution**
> > $\sqrt[3]{15} = \sqrt[3]{3 \times 5} = \sqrt[3]{3}\sqrt[3]{5}$ and $\sqrt[3]{75} = \sqrt[3]{3 \times 5 \times 5} = \sqrt[3]{3}\sqrt[3]{5}\sqrt[3]{5}$. Hence, $\sqrt[3]{9} \times 5$ is the most simplified form of $\sqrt[3]{15}\sqrt[3]{75}$.

7 Simplify the following radical expressions. Assume all variables are positive.

(a) $\sqrt{25a^3}$

(b) $\sqrt{20b^{10}}$

(c) $\sqrt{5x^2}\sqrt{5y^3}$

(d) $7\sqrt{2} \times 3\sqrt{y^2}$

(e) $\dfrac{\sqrt{35}}{\sqrt{7}}$

8 Assume that all variables are positive. Simplify.

(a) $\sqrt{48x^3}$

(b) $\sqrt[3]{216x^3y^4}$

(c) $\sqrt{75a^3}$

9 (Part 1. Multiplication) Assume that all variables are positive. Multiply and simplify.

(a) $\sqrt{3} \cdot \sqrt{8}$

(b) $\sqrt{ab} \cdot \sqrt{4ab}$

(c) $4\sqrt{3a^2} \cdot 2\sqrt{6a^3b}$

Other than multiplying radical expressions, we may have to divide the radicals. Especially when we divide two radicals, we simplify it by rationalizing the denominator. There are two types of radical quotients in Algebra 2.

- $\dfrac{\sqrt{a}}{\sqrt{b}}$ form : we multiply expressions to both numerator and denominator which make a denominator rational.

- $\dfrac{k}{\sqrt{a}-\sqrt{b}}$ form : we multiply the conjugate of $\sqrt{a}-\sqrt{b}$, which is $\sqrt{a}+\sqrt{b}$, to both numerator and denominator.

Example

Rationalize the denominator of $\sqrt{\dfrac{2x}{27y}}$. (Assume all variables are positive.)

Solution

$$\sqrt{\frac{2x}{27y}} = \frac{\sqrt{2x}}{\sqrt{27y}} = \frac{\sqrt{2x}\sqrt{3y}}{\sqrt{27y}\sqrt{3y}}. \text{ Then, } \frac{\sqrt{6xy}}{9y}.$$

---

9 (Part 2. Division) Simplify the following radical expressions. (Assume all variables are positive.)

(a) $\dfrac{\sqrt{4x}}{\sqrt{7y^2}}$

(b) $\dfrac{5}{\sqrt[3]{4x^2}}$

(c) $\dfrac{3\sqrt[3]{xy}}{\sqrt[3]{16x^2y}}$

(d) $\dfrac{\sqrt[4]{2x}}{\sqrt[4]{8x^3}}$

Given a binomial radical expression of form $\sqrt{a} \pm \sqrt{b}$, the simplification trick is to multiply conjugates. Before learning how to simplify a quotient expression, let's learn what *like radicals* are.

Example

Simplify $5\sqrt{3} + \sqrt{12}$.

> **Solution**
> At its first glance, it seems like it is impossible to add. However, simplifying $\sqrt{12} = 2\sqrt{3}$, you see that the radicands are equivalent. Hence, $5\sqrt{3} + 2\sqrt{3} = 7\sqrt{3}$.

---

10 Simplify, using the "like radicals."

(a) $5\sqrt{3} + \sqrt{18}$

(b) $4\sqrt{2} + \sqrt{8}$

One has to be careful when you see "simplification" inside a question. In fact, $\sqrt[4]{4}$ can be further simplified into $\sqrt[2]{2}$. If you feel bit challenged about simplifying radicals with other indices than 2, convert the given form into rational exponents and simplify it there.

11 Simplify, using the "like radicals."

(a) $5\sqrt[3]{3} + \sqrt{12}$

(b) $\sqrt[6]{8} + \sqrt{8}$

12 Simplify as much as possible.

(a) $\sqrt{28} - \sqrt{7}$

(b) $\sqrt{14} + \sqrt{35}$

13 Simplify the following radical quotients.

(a) $\dfrac{\sqrt{2}-\sqrt{5}}{\sqrt{3}-\sqrt{5}}$

(b) $\dfrac{4\sqrt{x}-\sqrt{3}}{2-\sqrt{7}}$

14 Simplify the following product. (In this problem, simply expand the given expressions.)

(a) $(\sqrt{3}-\sqrt{x})(\sqrt{3}+\sqrt{x})$

(b) $(\sqrt{8}+\sqrt{x})(\sqrt{x}-2\sqrt{2})$

7.3 Rational Exponents

Given $\sqrt[n]{x} = x^{\frac{1}{n}}$, we call the right hand side of the equation as the rational exponent expression of radicals. Exponential properties covered in Algebra 1 turn out to be valid in such expression, which allows us to expand the set of our inputs inside the exponents.

- $x^a x^b = x^{a+b}$

- $\dfrac{x^a}{x^b} = x^{a-b}$

- $(x^a)^b = x^{ab}$

The rules we have learned for exponential properties work perfectly fine with rational exponents. That being written, radical terms in the denominator with rational exponents, if any, may best be written in integer-exponent form, if possible, when being factorized.

15 Write the following expressions in simplest form. Assume all variables are positive.

(a) x^0

(b) x^{-1}

(c) $16^{\frac{3}{2}}$

(d) $(x^3 y^2)^{-\frac{1}{6}}$

(e) $\dfrac{x^{\frac{1}{2}} y^{\frac{1}{3}}}{x^{\frac{2}{3}} y^{\frac{-2}{3}}}$

(f) $\dfrac{x^{-4} y^{-1}}{x^2 y^{-3}}$

16 The rate of inflation $f(t)$ that raises the cost of an item from the present value P to the future value F over t years is given by the following formula $f(t) = \left(\dfrac{F}{P}\right)^{\frac{1}{t}} - 1$.

(a) What is the rate of inflation for which an Iphone that costs $1000 today will become $1500 in 3 years?

(b) What is the rate of inflation that will result in the price P doubling in 10 years?

17 Write the following expressions in simplest form. Assume that all variables are positive.

(a) $\left(\dfrac{x^{-\frac{1}{3}}y}{x^{\frac{2}{3}}y^{-\frac{1}{2}}}\right)^2$

(b) $\left(\dfrac{12x^8}{75y^{10}}\right)^{\frac{1}{2}}$

Solving Radical Equations

When we solve radical equations, our main goal is to isolate radical expression into either lefthand side or righthand side of the equation.

Example

Solve $\sqrt{x-2}-2=4$.

Solution

$\sqrt{x-2}=6$. Then, $x-2=36$. Hence, $x=38$.

18 Solve the following radical equations.

(a) $\sqrt[3]{2-x}=4$

(b) $(1+3x)^{\frac{1}{3}}=-2$

(c) $3x^{\frac{3}{2}}-1=5$

(d) $\sqrt{x}-\sqrt{x-1}=1$

19 Solve the following radical equations.

(a) $(3x+1)^{\frac{1}{3}} = 27$

(b) $3x^{\frac{4}{3}} + 4 = 52$

(c) $(x-1)^{\frac{2}{3}} - 4 = 5$

20 The formula $P = 4\sqrt{X}$ relates the perimeter P of a square to its area X. What is the area of the square whose perimeter is 24?

21 Solve the following radical equations. Check for extraneous solutions.

(a) $\sqrt{7x-6}-\sqrt{2x+4}=0$

(b) $\sqrt{s+9}-\sqrt{s}=1$

(c) $\sqrt{x+2}=x+2$

(d) $\sqrt{x}=x-6$

(e) $\sqrt{x+2}+18=x$

(f) $\sqrt{3-x}=2x-5$

7.5   Radical Function

Depending on the index, the graph of the function can be drawn in two forms.

- $y = \sqrt{x}$: the graph that is drawn in the first quadrant.

- $y = \sqrt[3]{x}$: the graph that is drawn in both first and third quadrant.

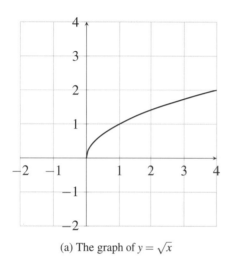

(a) The graph of $y = \sqrt{x}$

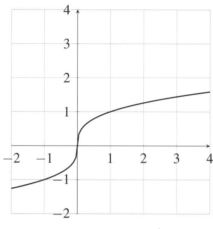

(b) The graph of $y = x^{\frac{1}{3}}$

Radical functions can be categorized into two types — even-indexed and odd-indexed. The family of even-indexed radical functions are similar to one another while different from the family of odd-indexed functions. Odd-indexed radicals are of form $\sqrt[3]{x}, \sqrt[5]{x}, \cdots$, while even-indexed radicals are of form $\sqrt{x}, \sqrt[4]{x}, \sqrt[6]{x}, \cdots$.

- **Even-indexed** Function : if a radical function is even-indexed, then **the domain depends on the expression inside the radicand** . Usually, the function $f(x) = \sqrt{x}$ is graphed only in the first quadrant.

- **Odd-indexed** Function : if a radical function is odd-indexed, then **the domain is the set of all real numbers**. The function $f(x) = \sqrt[3]{x}$ is graphed overall, most of which is drawn at the first or third quadrant.

If we summarize the domain and range of radical functions, we get

- Domain : it depends on the parity of the index.

    Even index : the radicand ≥ 0

    Odd index : $\mathbb{R}$

- Range : it depends on the parity of the index.

    Even index : the radical expression ≥ 0

    Odd index : $\mathbb{R}$.

22 Graph $y = \sqrt{x} + 1$.

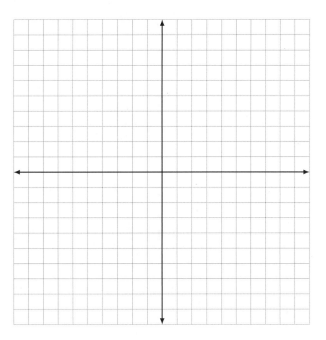

23 Graph $y = \sqrt{2x}$.

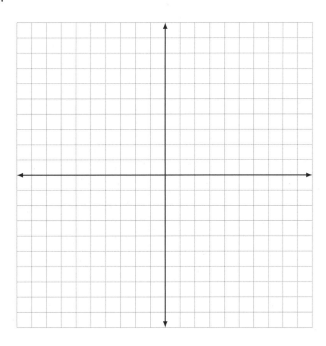

24 Graph $y = -\sqrt{-x+2} + 3$.

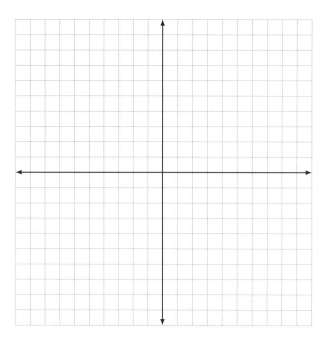

25 Graph $y = 2\sqrt{x+3} - 3$.

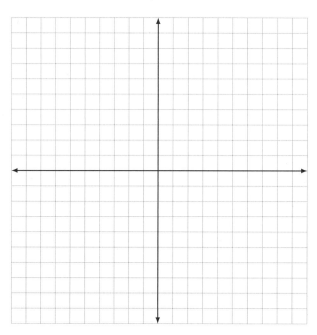

1. If $0 < x < 1$, then simplify $\sqrt{\left(1+\dfrac{1}{x}\right)^2} - \sqrt{\left(1-\dfrac{1}{x}\right)^2}$.

2. Simplify $\sqrt{12} - \dfrac{9}{\sqrt{27}}$.

3. If $\dfrac{4}{\sqrt{3}+1}$ is simplified to $A + B\sqrt{3}$ for rational numbers A and B, then find $B - A$.

4 Simplify the following expressions.

(a) $\sqrt{18} - 3\sqrt{2}$

(b) $4\sqrt{2} - \sqrt{8} + \sqrt{48} - 3\sqrt{3}$

(c) $\sqrt{48} + 5\sqrt{3} - \dfrac{6}{\sqrt{3}}$

5 If $\sqrt{150} = a\sqrt{6}$ and $\sqrt{\dfrac{32}{9}} = b\sqrt{2}$ for rational numbers a and b, then find $a+b$.

6 If $\dfrac{\sqrt{12}-2}{\sqrt{3}-2} - \dfrac{2}{\sqrt{5}+\sqrt{3}} = a+b\sqrt{3}+c\sqrt{5}$ for rational numbers a, b, and c, then find $a+b+c$.

7 Solve $\sqrt{6x+1}-1=2x$.

8 Solve $\sqrt{4m-3}-2=\sqrt{2m-5}$.

9 Solve $\sqrt{x+1}+\sqrt{x}=1$.

10 Solve $\dfrac{\sqrt{x-2}}{x-2}=\dfrac{x-5}{\sqrt{x-2}}$.

11 Solve $\sqrt{\sqrt{x}+5}=3$.

12 Solve $\sqrt{x^2 + 6x} - 2\sqrt{6} = 0$.

13 Solve $\sqrt[3]{x-2} + 4 = 2$.

14 Solve $\sqrt{2x+5} + \sqrt{x+2} = 5$.

15 Graph $f(x) = \sqrt{x-3} + 2$.

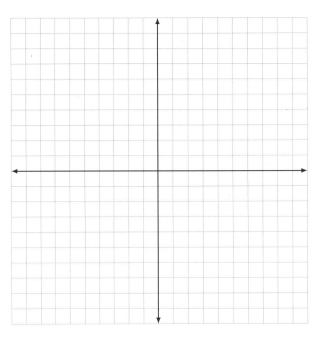

16 Graph $f(x) = -\sqrt{x-3} + 1$.

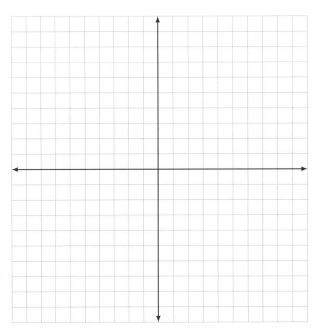

17 Graph $f(x) = \sqrt{3-x} + 1$.

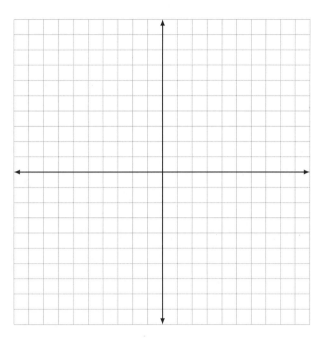

18 Graph $f(x) = 2\sqrt{x} - 2$.

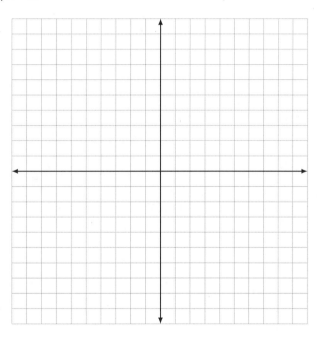

19 Graph $f(x) = -\sqrt{-x+1} - 1$.

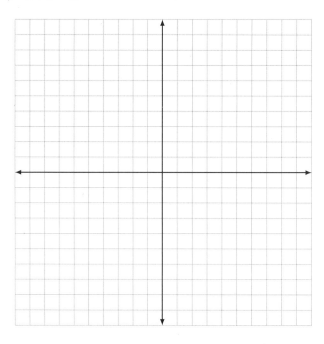

20 Graph $f(x) = \sqrt[3]{x+2} + 3$.

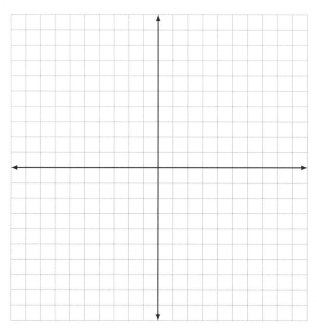

1 2

2 $\sqrt{3}$

3 4

4

(a) 0   (b) $2\sqrt{2}+\sqrt{3}$ (c) $7\sqrt{3}$

5 $a+b=\dfrac{19}{3}$

6 $a+b+c=-4$

7 $x=0,\dfrac{1}{2}$

8 $m=3,7$

9 $x=0$

10 $x=6$

11 $x=16$

12 $x=-3\pm\sqrt{33}$

13 $x=-6$

14 $x=2$

15

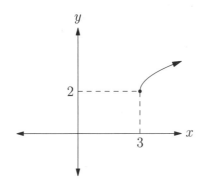

16

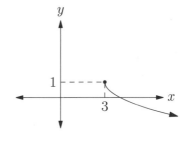

17

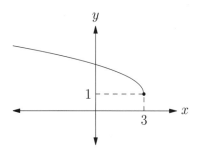

18

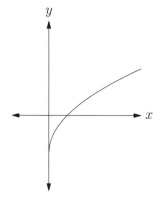

19

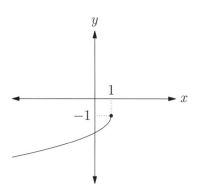

20

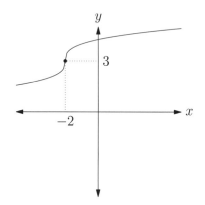

Topic 8

Rational Expression and

Rational Function

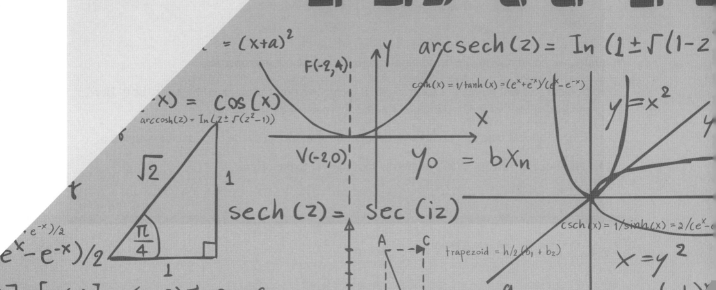

8.1 Inverse Variation

When we say a variable x varies inversely with y, we get the equation of the form

$$x = \frac{k}{y}$$

where k is the constant of variation. Unlike direct variation ($y = kx$ for some constant k), inverse variation has a rational form.

Example

x varies inversely with the square root of y. If $x = 3$, then $y = 16$. Find x if $y = 9$.

> **Solution**
> Since this is the inverse variation, $x = \frac{k}{\sqrt{y}}$. Then, $k = 12$. Hence, $\frac{12}{\sqrt{9}} = 4 = x$.

The following diagram is the graph of the parent function $y = \frac{1}{x}$, which is the inverse variation between x and y with the constant of variation as 1.

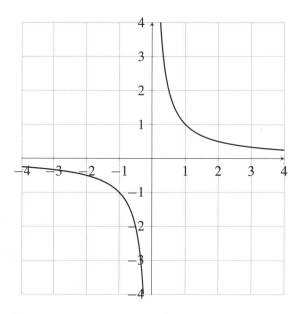

Naturally, we could investigate its domain, range, and some other properties.

- Domain : all real numbers except $x = 0$. It means we don't want denominator to be 0. Here, $x = 0$ is called "vertical asymptote" (a line approached by the curve).

- Range : all real numbers except $y = 0$, which is also known as the horizontal asymptote.

- Odd function : the graph is symmetric with respect to the origin.

1 Each pair of values forms a direct variation. Find the missing value.

(a) $(2,6)$ and $(3,m)$

(b) $(3,-6)$ and $(-1,t)$

(c) $(4,3)$ and $(x,-6)$

2 Assume y varies directly as x.

(a) If $y=4$ when $x=7$, find x when $y=10$.

(b) If $y=12$ when $x=-3$, find y when $x=4$.

3 Each pair of values forms an inverse variation. Find the missing value.

(a) $(2,6)$ and $(3,k)$

(b) $(3,-6)$ and $(-1,x)$

(c) $(4,3)$ and $(x,-6)$

4

(a) Suppose y varies inversely with the square of x, and $y=100$ when $x=4$. Find y when $x=10$.

(b) Suppose x varies jointly[1] with y and z, and $x=10$ when $y=2$ and $z=-10$. Find x when $y=5$ and $z=20$.

---

[1] Joint variation is covered in Algebra 1. If a is jointly varying with b and c, we can write $a=kbc$ for some constant of variation k.

8.2   Graphs of Rational Functions

Given the graph of $y = \dfrac{1}{x}$, we can apply the usual transformation of shifting, scaling, and reflecting. However, there is another way of thinking about the graph. Since $x \neq 0$, we multiply it to the both sides of the equation. In that case, we get $xy = 1$, which looks similar to the area formula for the rectangle. As long as $x \neq 0$ and $y \neq 0$, we get two possible cases : $(+) \times (+) = 1$ or $(-) \times (-) = 1$. For the first case, the graph stays on northeast portion of the regions cut by horizontal and vertical asymptotes. Similarly, for the second case, the graph stays on their southwest portion.

Example

Describe the transformation from $y = \dfrac{1}{x}$ into $y = \dfrac{1}{2-x} + 3$.

Solution

First, $y = \dfrac{1}{-(x-2)} + 3$ means that the graph of $y = \dfrac{1}{x}$ is reflected about the y-axis, and it is shifted 2 unit right and 3 unit up. Hence, the vertical asymptote is $x = 2$ and the horizontal asymptote is $y = 3$.

$\boxed{5}$ Sketch the following rational graphs.

(a) $f(x) = \dfrac{1}{x-1} + 1$.

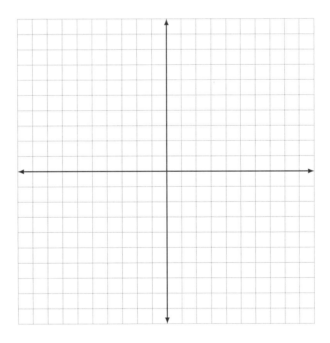

Describe the graph of $y = \dfrac{1}{x-1}$ in terms of area method.

Solution

Notice that $y(x-1) = 1$. Consider $x - 1$ as the base of a rectangle and y as the height. It is obvious that $x \neq 1$ and $y \neq 0$. Otherwise, the lefthand side of the equation will be 0. Now, let's think about all possible cases that start with one of the vertex of the rectangle located at $(1, 0)$. If you move one unit right from $x = 1$ and one unit up from $y = 0$, we get the area of the rectangle (in this case, a square) as 1. Check that $(2, 1)$ is a solution to the given rational expression. Likewise, if you move two units right from $x = 1$ and one-half unit from $y = 0$, we still get the area of the rectangle as 1. Check that $(3, 1/2)$ is also part of the solution set. How about moving to the leftside of $(1, 0)$? If you move one unit left from $x = 1$, then one unit down from $y = 0$ to get the area of rectangle as 1 because of $(-1)(-1) = 1$. Hence, we get $(0, -1)$ in the solution set. Either way, we need to make signed area of 1 whose vertex is fixed at $(1, 0)$.

Try graphing the following example with the perspective laid out in the example above. First, convert the following example into the form $y - 2 = \dfrac{1}{2-x}$, or $y - 2 = \dfrac{-1}{x-2}$. In this case, $(y-2)(x-2) = -1$, which resembles an area formula for rectangle with the signed area of -1. In other words, we need to have $(-)(+) = -1$ or $(+)(-) = -1$. Make sure you label the vertical asymptote at $x = 2$ and horizontal asymptote at $y = 2$.

(b) $f(x) = \dfrac{1}{2-x} + 2$.

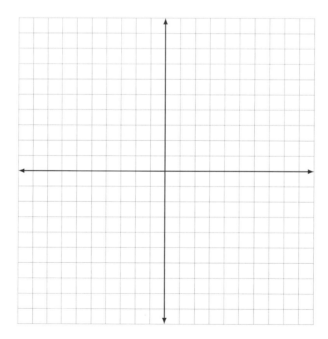

8.3 Reciprocal Function

Other than transformation or rectangle-area method covered in the previous section, the study of rational function also uses the property of reciprocal function. Given $y = f(x)$, the reciprocal function $y = \dfrac{1}{f(x)}$ satisfies the following properties.

- Sign of y-value does not change.

- Vertical asymptote becomes x-intercept, vice versa.

- Whatever goes to $\pm\infty$ becomes a horizontal asymptote.

- Increasing part $\longleftrightarrow$ Decreasing part

$\boxed{6}$ Sketch the graph of the following functions.

(a) $g(x) = -\dfrac{1}{x^2}$.

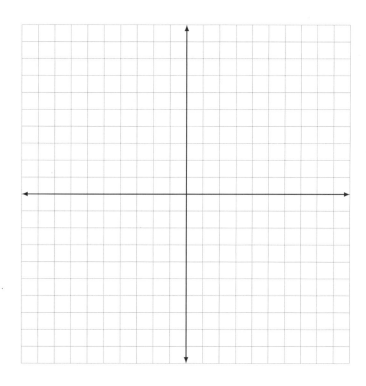

Let's go over an example of making a quadratic reciprocal function. First, sketch the graph of $y = x^2 + 3x - 4$. Look at its x-intercepts. They will all turn into the vertical asymptotes. Now, look at which parts of the graph have positive y-values or negative y-values. The sign of y-values will not change, meaning that respective portions of the graph will be above or below the x-axis, in accordance with the signs of y-value of the original graph. The figure below shows the graph of $y = x^2 + 3x - 4$. Use it to solve for part (b).

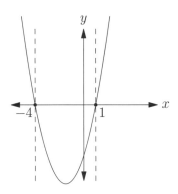

(b) $h(x) = \dfrac{1}{x^2 + 3x - 4}$

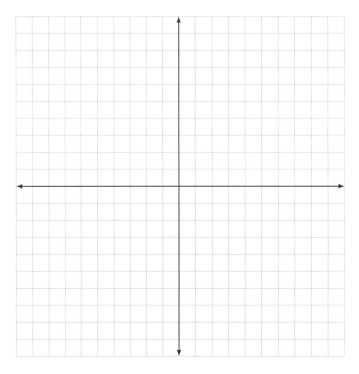

8.4 Hole or Vertical Asymptote

Given $y = \dfrac{p(x)}{q(x)}$, we look at denominator carefully to see which value of x is *undefined*. Out of these undefined values, we can categorize them into two types.

- Hole : the term both in *denominator* and *numerator* that is canceled when simplified.

- Vertical asymptote : the term in *denominator* that is left when simplified.

Example

Find the hole or vertical asymptote of $y = \dfrac{(x-1)(x-2)}{(x-2)(x-3)}$.

Solution
The domain is all real numbers except $x = 2$ and $x = 3$. Here, $(x-2)$ is a common term in both numerator and denominator. Hence, $x = 2$ is a hole. On the other hand, $(x-3)$ is left after simplification. Hence, $x = 3$ is the vertical asymptote.

---

7 Find the hole, if any, of the following functions.

(a) $y = \dfrac{x-3}{x^2-9}$

(b) $y = \dfrac{x^2-1}{x-1}$

(c) $y = \dfrac{x^2+3x+2}{3x^2+4x+1}$

8.5 Rational Inequality

When we solve rational inequality, we use two methods. The first method is sign analysis that takes some caseworks. The other method is a quicker method that uses polynomial forms.

Example

Solve the following inequality $\dfrac{x-1}{x-2} < 0$.

Solution

The first method is sign analysis. We look at all possible cases.

If $x < 1$, the $x - 1 < 0$ and $x - 2 < 0$. Therefore, $\frac{x-1}{x-2} > 0$. This can't be right. If $x = 1$, then $\frac{0}{x-2} = 0$, so this is also wrong. If $1 < x < 2$, then $x - 1 > 0$ but $x - 2 < 0$. Voila! This works. If $x = 2$, then $\frac{x-1}{x-2}$ is undefined. If $2 < x$, then $x - 1 > 0$ and $x - 2 > 0$. Hence, it does not work. So, we are left with the only interval $(1,2)$.

The second method is multiplying $(x-2)^2$ to both hand sides of the inequality. Since $(x-2)^2 > 0$ for $x \neq 2$, the inequality does not change. Hence, we get $(x-2)(x-1) < 0$. If you solve this quadratic inequality, you get $(1,2)$ as well.

8

(a) Solve the following inequality $\dfrac{x-2}{x+1} \leq 0$ by sign analysis.

(b) Sketch the graph of $y = \dfrac{x-2}{x+1}$ and see if the solution in part (a) matches with the graph.

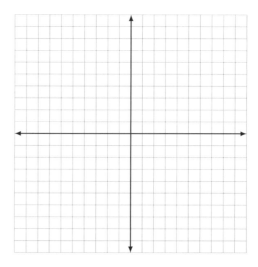

8.6 Word Problems

There are two types of word problems using rational equations in algebra 2, most of which use harmonic mean of two variables[2].

- Work-rate question : given a number of people(or machines) and their workrate, we figure out how long it takes to finish a work in different number of people(or machines).

- Speed-rate question : we use the notion of $d = rt$ where d is the distance traveled, r is the speed of the object of our interest, and t is the time it takes for the object to travel d distance.

Example

Ben can paint a house in 5 hours. If he works with Lauren, two can finish painting the house in 2 hours. How long would it take for Lauren to finish painting the house on her own?

Solution

In an hour, Ben can paint $\dfrac{1}{5}$ of the house. If the number of hours for Lauren to finish painting the house is x, then she can paint $\dfrac{1}{x}$ of the house in an hour. Therefore, $\left(\dfrac{1}{5} + \dfrac{1}{x}\right) 2 = 1$. Hence, $x = \dfrac{10}{3}$ hours.

9 If two water pipes can fill a pool in two hours, and one pipe can fill the pool in four hours, how long would the other pipe to fill the pool on its own?

10 Bob drove 50 miles from his office to his house at an average speed of 100 miles per hour. Going to his office, he encountered heavy traffic and drove the same 50 miles at an average speed of 50 miles per hour. What was his average speed for the entire 100-mile round trip? (All of this Bob situation is hypothetical. You should not drive at an average speed of 100 miles per hour. Be safe.)

---

[2]Given two real numbers a and b, the harmonic mean equals $\dfrac{2}{\frac{1}{a} + \frac{1}{b}}$.

8.7 Solving Rational Equations

There are three steps we use when we solve fractional equations.

- Equalize the denominators, if possible.

- Combine multiple expressions into one expression.

- Get rid of values that make denominator 0 because they are extraneous solutions.

Example

Solve $\dfrac{3}{2} - \dfrac{1}{x} = \dfrac{5}{6}$.

Solution

Since $\dfrac{3}{2} - \dfrac{1}{x} = \dfrac{3x-2}{2x}$, we get $\dfrac{3x-2}{2x} = \dfrac{5}{6}$. Hence, $18x - 12 = 10x$. Therefore, $8x = 12$, so $x = 1.5$.

11 Solve $\dfrac{1}{x-3} + \dfrac{1}{x+3} = \dfrac{4}{x^2-9}$.

12 Solve $\dfrac{1}{2x} - \dfrac{1}{3(x+5)} = \dfrac{1}{x}$.

1 Sketch the graph of $y = \dfrac{2x+4}{x-1}$.

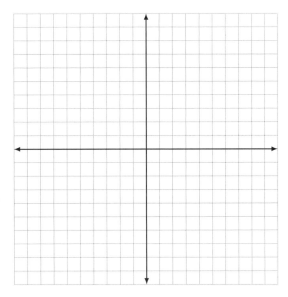

2 Sketch the graph of $y = \dfrac{-2x+2}{2x+4}$.

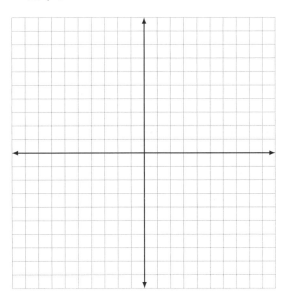

3 Sketch the graph of $y = \dfrac{4}{x^2 + 1}$.

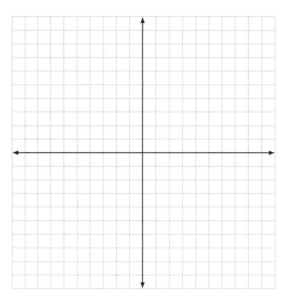

4 Sketch the graph of $y = \dfrac{1}{(x-1)(x+1)}$.

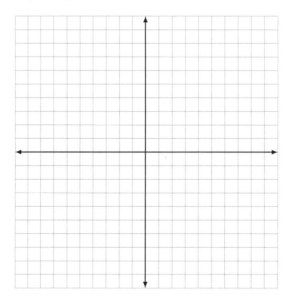

5 Sketch the graph of $y = \dfrac{x+1}{x^2-1}$.

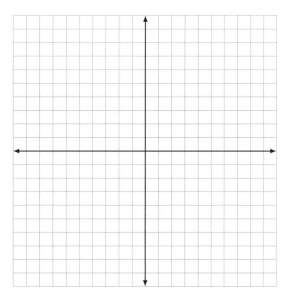

6 Sketch the graph of $y = \dfrac{x^2-9}{x^2-2x-3}$.

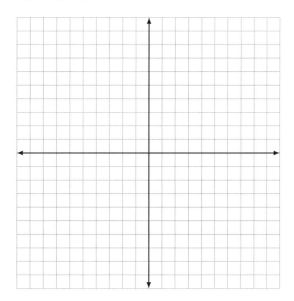

7 Sketch the graph of $y = \dfrac{x}{x^2 - x - 2}$.

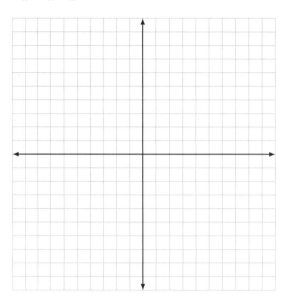

8 Sketch the graph of $y = \dfrac{x^2 - x - 6}{x - 3}$.

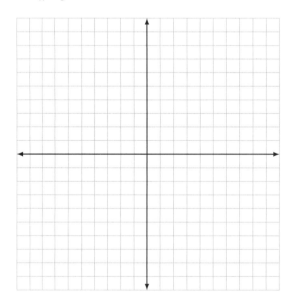

9 Find the vertical asymptote of the following rational functions.

(a) $f(x) = \dfrac{2+x}{3+x}$

(b) $g(x) = \dfrac{2-x}{x^2-4}$

(c) $h(x) = \dfrac{2+3x}{4+6x}$

10 Find the horizontal asymptote of the following rational functions.

(a) $f(x) = \dfrac{2+x^2}{3x+4x^3}$

(b) $g(x) = \dfrac{3-2x}{5x+7}$

(c) $h(x) = \dfrac{3x^3+4x-1}{2x^2-8}$

11 Find the oblique[3] asymptote of a rational function $f(x) = \dfrac{4x^3 + 2x^2 - 3}{x^2 - 3}$.

12 Solve the inequality $\dfrac{(x-1)^2}{x+3} > 0$.

13 Solve the inequality $\dfrac{3x+2}{(x-1)(x+2)} \leq 0$.

---

[3]Oblique asymptote refers to $y = mx + b$ which is equivalent to the quotient for $y = \frac{p(x)}{q(x)}$. Normally, if the quotient is a constant k, then we call it $y = k$, a horizontal asymptote.

14 The faster train takes 2 hours shorter to travel 150 miles than the slower train to travel 170 miles. If the difference of the average speed is 16 miles per hour, find the speed of the slower train.

15 Bob has 10 liters of a juice blend that is 50% juice. How many liters of pure juice should he put to make it 70%?

16 Solve for x.

(a) $\dfrac{2}{x+1} + \dfrac{x}{x-1} = \dfrac{2}{x^2-1}$

(b) $\dfrac{x}{x-1} + \dfrac{-2}{x+1} = \dfrac{2}{3}$

1

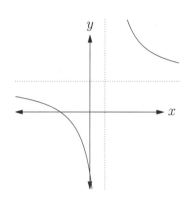

2

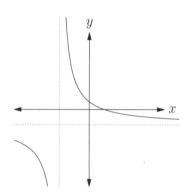

3

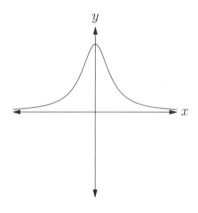

4

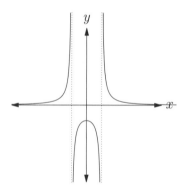

5

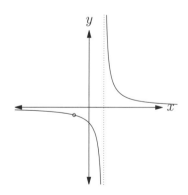

6

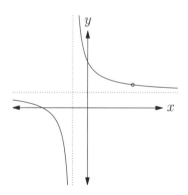

7

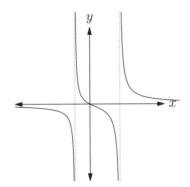

8

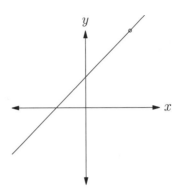

9

(a) $x = -3$ (b) $x = -2$ (c) None

10

(a) $y = 0$ (b) $y = -\dfrac{2}{5}$ (c) None

11 The oblique(slant) asymptote is $y = 4x + 2$.

12 $-3 < x < 1$ or $1 < x$.

13 $x < -2$ or $-\dfrac{2}{3} \leq x < 1$.

14 34 mph.

15 $\dfrac{20}{3}$ liters of juice.

16

(a) $x = -4$.

(b) no real solution.
(In fact, $x = \dfrac{3 \pm \sqrt{23}i}{2}$ are complex solutions.)

Algebra 2와 수학 경시대회 관련한 Q&A

Q. AMC 10/12 또는 9, 10학년이 출전할 수 있는 경시대회 준비를 위해, Algebra 2는 반드시 학습해야 하나요?

A. AMC 10/12 공통 주제로 Algebra 2에서 다루는 개념들을 활용한 문제들은 자주 출제되는 편입니다. 예를 들어, rational function을 활용한 AM-GM inequality는 자주 출제된다고는 할 수 없지만 출제 빈도수가 적은 편도 아니기 때문에, 반드시 알아야 하는 내용입니다. 특히, counting and probability 부분은 AMC 10/12 뿐 아니라, AMC 8에서부터 꾸준히 출제되고 있는 개념 부문이므로, 개념을 한번 잡고 가는 것이 필수입니다.

Q. 이 교재로 수학 경시대회 준비가 가능할까요?

A. 이 교재는 Algebra 2를 처음 배우고자 하는 학생들을 위해, 개념 입문편이라고 볼 수 있습니다. 개념을 한번도 학습해보지 않은 학생들에게는 추천하는 교재입니다. 즉, Algebra 2를 한번도 수강하지 않은 학생들이라면, 이 교재를 통해 먼저 학습하는 것을 추천합니다. 다만, 수학 경시대회 대비를 위한 교재로 활용하기에는 응용의 폭 및 깊이 부분에서 부족하니, 저자의 'Essential Math Serie's 중 〈The Essential Guide to Competition Math : Counting and Probability〉를 선택하여 학습하는 것이 오히려 효과적일 수 있습니다.

Q. 한국 수학 커리큘럼상 수1, 2 개념을 학습한 상태면, Algebra 2 개념 학습 없이도, 수학 경시대회 준비가 바로 가능할까요?

A. 이 부분은 학생들마다 다르게 적용합니다. 예를 들어, 수1, 2 개념을 학습하여서, Algebra 2에 있는 개념들이 대부분 학습된 상태일 때, 수학 경시대회 준비는 가능하지만, Algebra 2 개념을 학습할 때 배우는 특정한 방식들에 대해서 노출되지 않았다면, 개념에 공백이 있을 수 있습니다. 이러한 이유로, 학생들 개별 상황에 맞추어서 대비하는 것이 좋습니다.

Topic 9

Exponential and Logarithmic

Expressions and Functions

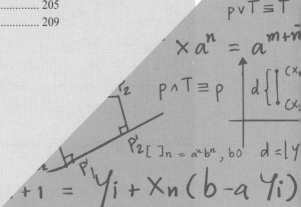

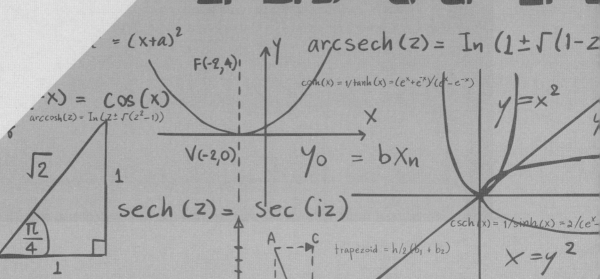

9.1 Exponential Function

There are two types of basic exponential functions.

- Increasing function : $y = b^x$, where $1 < b$

- Decreasing function : $y = b^x$, where $0 < b < 1$

The exponential function $y = b^x$ is either *monotonically increasing* or *decreasing* function, which means it is $1-1$ function. If a function is 1-to-1, it means that if y-values are different, then x-values are different. In other words, y-values are redundant.

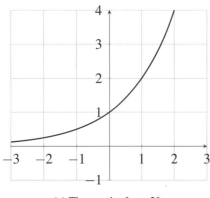

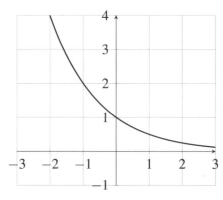

(a) The graph of $y = 2^x$ (b) The graph of $y = 2^{-x}$

$\boxed{1}$ Sketch the graph of the following exponential function $f(x) = 2^{x-1} + 1$.

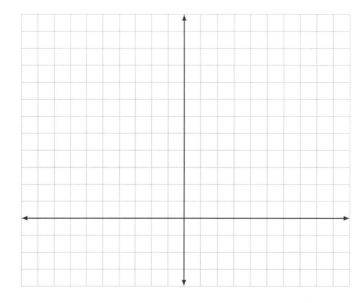

$\boxed{2}$ A function $f(x) = 3^x - 2$ is an exponential function.

(a) Determine the domain.

(b) Determine the range.

(c) Determine the horizontal asymptote, if any.

$\boxed{3}$ Sketch $f(x) = -2^{3-x}$ and state the domain, range, and horizontal asymptote.

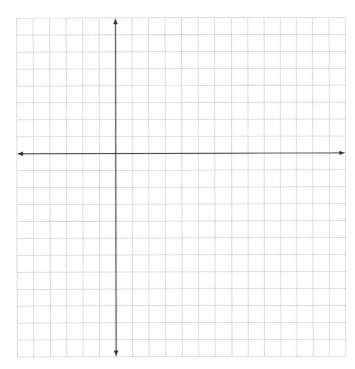

9.2 Exponential Equation

Along with exponential expressions in Algebra 1, exponential equation uses a^m where $a > 0$. The following list shows the properties of indices. Assume that $a > 0$, and p, q are positive integers.

- $a^0 = 1$

- $a^{-p} = \dfrac{1}{a^p}$

- $a^{\frac{1}{p}} = \sqrt[p]{a}$

- $a^{\frac{q}{p}} = (\sqrt[p]{a})^q$

The following arithmetic rules for indices can easily be deduced and they are quite handy. Memorize them by heart. Assume $a, b > 0$ and m and n are rational numbers.

- $a^m \cdot a^n = a^{m+n}$

- $a^n \cdot b^n = (ab)^n$

- $\dfrac{a^m}{a^n} = a^{m-n}$

- $\dfrac{a^n}{b^n} = (\dfrac{a}{b})^n$

- $(a^m)^n = a^{mn}$

$\boxed{4}$ Solve the following exponential equation. (Hint: Set bases equal.)

(a) $4^{2x+2} = 8$

(b) $3^{x+2} = \dfrac{1}{27}$

(c) $3(9)^{x+3} = \dfrac{1}{27^x}$

5 Solve for x.

(a) $9(3)^x = 27^{1-x}$

(b) $16^{x-1} = \dfrac{\sqrt{2}}{8^{-x}}$

(c) $\dfrac{2^{x+2}}{16} = 4^{2x-1}$

(d) $9^{\sqrt{x+1}} = \dfrac{1}{3^{7-x}}$

(e) $4^{x^2+2x} = \dfrac{1}{64}$

6 Solve for x.

(a) $3^{2x} - 4(3^x) + 3 = 0$

(b) $2^{2x} - 2^{x+1} - 8 = 0$

(c) $5^{2x} + 25 = 26(5^x)$

(d) $3^{1+x} = \dfrac{9}{3^x} + 26$

(e) $2^{2x} + 2^{x+2} = 8$

There is a mathematical model using exponential growth or decay - compound interest rates or population growth. If r is the annual interest rate written in decimal expression[1], and P_0 is the principal(original) amount in the bank account, the savings account after n years, without any withdrawal, may be modeled under compound interest rate as

$$P_0(1+r)^n = P(n)$$

Same tool may be applied to population growth. If r is the annual growth rate, and P_0 is the original population at $t = 0$, the number of population after n years of exponential growth will be modeled by

$$P_0(1+r)^t = P(t)$$

In the next three practice problems, assume that we stick to exponential growth model.

$\boxed{7}$ Suppose you deposit $1,500$ dollars in a savings account that pays an annual interest rate of 4%. Suppose no money is added or withdrawn from the account, how much will be in the account after 4 years?

$\boxed{8}$ A population of $130,000$ grows 2% per year for 15 years. How much will the population be after 15 years?

$\boxed{9}$ The initial value of a car is $72,000$. After one year, the value dropped to $57,600$. What exponential function could model the expected value of a car after three years?

---

[1] If $r = 60\%$, we generally write $r = 0.6$.

9.3 Logarithmic Function

Given an exponential form $a^x = b$, we have the logarithmic form, which is equivalent to the exponential one, i.e.,

$$\log_a(b) = x$$

where $b > 0$ and x can be any real number. The graph of logarithmic function also depends on the base value.

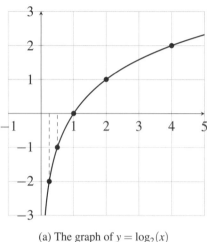

(a) The graph of $y = \log_2(x)$

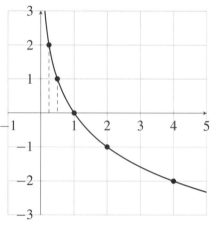

(b) The graph of $y = \log_{\frac{1}{2}}(x)$

10 Graph $y = \log_3(x)$ in the following xy-plane with proper labels.

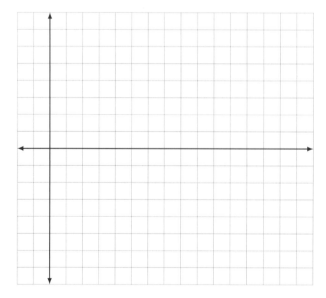

9.4 Logarithmic Equation

The following lists illustrate the properties of logarithms.

Logarithmic Properties

1. $\log_a b^n = n \log_a b$

2. $\log_a b + \log_a c = \log_a bc$

3. $\log_a b - \log_a c = \log_a \dfrac{b}{c}$

4. $(\log_a b) \times (\log_c d) = (\log_a d)(\log_c d)$

5. $\dfrac{\log_a b}{\log_a c} = \log_c b$

6. $\log_{a^n} b^m = \dfrac{m}{n} \log_a b$

There are two common logarithms used over the courseworks in high school: common logarithm and natural logarithm. These two logarithms have a unique notation without writing the base.

Common Logarithm and Natural Logarithm

1. $\log_{10} x = \log(x) = \lg(x)$

2. $\log_e x = \ln(x)$

Here, e refers to Euler number, which is slightly bigger than 2. Definition of e involves a limit process, which will not be covered in Algebra 2. Nevertheless, knowing that e is bigger than 2 is quite useful when we graph the function $y = \ln(x)$.

$\boxed{11}$ Express each of the following exponential expressions in logarithmic form.

(a) $2^5 = 32$

(b) $2^{-1} = \dfrac{1}{2}$

(c) $4 = (x+1)^2$

(d) $a^{-m} = b$

$\boxed{12}$ Express each of the following logarithmic expressions in exponential form.

(a) $\log_2 16 = 4$

(b) $\log_{\frac{3}{2}} \dfrac{9}{4} = 2$

(c) $\log_{27} 81 = \dfrac{4}{3}$

13 Evaluate the following logarithmic expressions.

(a) $\log_2 8$ (b) $\log_3 27$ (c) $\log_3 3$ (d) $\log_4 1$

14 Solve for x.

(a) $\log_{\sqrt{2}} x = 2$ (b) $\log_x(\sqrt{2}) = 4$ (c) $\log_{x-1}\sqrt{5} = \dfrac{1}{2}$

15 Write each of the following as a single logarithm.

(a) $\log_3 9 + \log_3 27$ (b) $\log_3 8 - 1$ (c) $2\log_2 3 + 3\log_2 9$

16 Without using a calculator, evaluate the following expressions.

(a) $3^{\log_3(x)}$

(b) $e^{\ln(4)}$

(c) $2^{2\log_2 4}$

17 Simplify the following expressions.

(a) $\log_2 3 \times \log_3 5 \times \log_5 4$

(b) $e^{\ln(3)} \times 5^{2\log_{\sqrt{5}} 4}$

(c) $\log_{125} 5 + \log_{27} 9$

(d) $\dfrac{\log_2 25}{\log_2 5} \times \dfrac{\log(16)}{\log(4)}$

18 Evaluate each of the following in terms of x and y given $x = \log_2 3$ and $y = \log_2 5$.

(a) $\log_2 15$

(d) $\log_3 15$

(b) $\log_2(7.5)$

(e) $\log_4 9$

(c) $\log_3 2$

(f) $\log_5 6$

9.5 Using Logarithm to Solve Equations

Given $a^x = b$ where $a > 0$, except $a = 1$, and $b > 0$, we can find a solution x by using two procedures.

1. If b can be written as a^n, then
$$b = a^x = a^n \implies x = n$$

2. Otherwise, take logarithms on both sides, i.e.,
$$\log a^x = \log b \implies x = \frac{\log b}{\log a}$$

In order to decide when to use logarithm to solve a given equation, first look at whether the exponent contains a variable. If so, logarithm might take down the exponent to a constant multiple and make a linear equation, which is easier to solve. If expressions have common exponents, put all bases inside one base in order to make the expressions in the following form.

$$\bigcirc^{\square} = \triangle$$

Also, the wise choice of base might be helpful in simplifying an exponential equation. For instance, if $3^x = 10$, then it is better to take $\log_3(3^x) = \log_3(10)$ rather than $\log_5(3^x) = \log_5(10)$ (or any other base). Nevertheless, the nature of a solution retrieved from any choice of base does not change. In fact, it is simply rewriting answer in a different form.

$$x = \log_3(10) = \frac{\log_5(10)}{\log_5(3)}$$

19 Solve for x.

(a) $3^x - 4 = 7$ (b) $5^{2x} - 3 = 4$ (c) $100e^{2x} = 200$

20 Solve for x.

(a) $2^{2x} - 4(2^x) = 5$ (b) $3^x - 3(3^{-x}) - 2 = 0$

21 Solve for x.

(a) $3^{2x} = 10^{1-x}$

(b) $4^{x-1} = 3^{2+x}$

(c) $\dfrac{m^{2x-1}}{n^x} = p^{2-x}$

22 Solve for x.

(a) $\log_2(3x) = 8$

(b) $\log(2x+3) = -2$

(c) $\log_3(1+x^2) = 27$

23 Solve for x.

(a) $\log_3(\log_2(x)) = 1$

(b) $\log_4(\log_{27}(x-1)) = \dfrac{1}{2}$

24 Solve for x.

(a) $\log_5(5x+40) - \log_5(x+2) = 2$

(b) $\log_3(3-2x) + \log_3(x+1) = 1$

(c) $\ln(x-9) + \ln(x) = \ln(10)$

(d) $\log_2(x-1) - \log_4(x+2) = 1$

$\boxed{1}$ Given the function $f(x) = 2 \times 3^x$ for $-2 \le x \le 4$,

(a) find the range of $f(x)$.

(b) find the value of x given that $f(x) = 162$.

$\boxed{2}$ Given a "function" $y = a^x$, if it passes through $(4, 16)$, find the y-intercept.

3 Given an exponential function $y = 3^{-x} + 2$, the curve passes through $(0, m)$ and $(1, n)$.

(a) Find the value of m and n.

(b) Find the asymptote of the curve.

4 Given a function $f(x) = \dfrac{5}{4} - a^{-x}$, if $(2, 1)$ is on the graph of $y = f(x)$, then find the value of a.

5 Solve the equation $2^{2x+2} - 10 \cdot 2^x + 4 = 0$ for x where $x \in \mathbb{R}$.

6 Without using a calculator, solve for x.

(a) $27^{-x} = 3$

(b) $9^{x-1} = 27^{-x-2}$

7 Solve the following system of equations.

$$\begin{cases} 4^x \times 2^y = 16 \\ 8^x = 2^{\frac{y}{2}} \end{cases}$$

8 Solve for x.

(a) $6 \times 2^x = 192$

(b) $4 \times \left(\frac{1}{3}\right)^x = 324$

9 Solve $e^{-x^2+3} = (e^2)^x$ for x.

10 Find the amount in the account satisfying the following conditions. Assume that no money is withdrawn in the account during the given time period.

(a) Compound interest rate, principal : $1000, annual interest rate : 3%, time : 2 years.

(b) Continuously[2] compounded, principal : $1000, annual interest rate : 3%, time : 2 years.

---

[2]A model for continuously compounded interest rate when the annual interest rate is r, in decimal expressions, with the principal amount of P_0, is given by $P_0 e^{rt}$ for t in years.

11 If $\log 2 = m$, express $\log_8 5$ in terms of m.

12 Solve the equation $(\log_3 y)^2 + \log_3(y^2) = 8$.

13 Express $\log_x 2$ in terms of a logarithm to base 2. Using this expression, find the values of x which satisfy the equation $\log_2 x = 3 - 2\log_x 2$.

14 Find the value of $2\log_{12} 4 - \dfrac{1}{2}\log_{12} 81 + 4\log_{12} 3$.

15 If $\log p^3 q = 10a$ and $\log \dfrac{p}{q^2} = a$, find, in terms of a, expressions for $\log p$ and $\log q$. Also, find the value of $\log_p q$.

16 Given that $\log_9 x = a \log_3 x$, find a. Given that $\log_{27} y = b \log_3 y$, find b. Hence, solve for x and y for the simultaneous equations

$$6 \log_9 x + 3 \log_{27} y = 8$$
$$\log_3 x + 2 \log_9 y = 2$$

17 Solve the equation

(a) $2^{2x+1} = 20$

(b) $\dfrac{5^{4y-1}}{25^y} = \dfrac{125^{y+3}}{25^{2-y}}$

18 Solve the equation.

(a) $\log y + \log(y - 15) = 2$

(b) $\dfrac{5^{2x+3}}{25^{2x}} = \dfrac{25^{2-x}}{125^x}$

19 Given that $u = \log_4 x$, find, in simplest form in terms of u.

(a) x

(b) $\log_4\left(\dfrac{16}{x}\right)$

(c) $\log_x 8$

1

(a) $[\frac{2}{9}, 162]$

(b) $x = 4$

2 The y-intercept is 1.

3

(a) $m = 3$, $n = 2\frac{1}{3}$

(b) $y = 2$

4 $a = 2$

5 $x = -1, 1$

6

(a) $x = -\frac{1}{3}$

(b) $x = -\frac{4}{5}$

7 $(x, y) = \left(\frac{1}{2}, 3\right)$

8

(a) $x = 5$

(b) $x = -4$

9 $x = -3, 1$

10

(a) $1000 \times (1.03)^2 \approx 1060.9$ dollars

(b) $1000e^{0.06} \approx 1061.84$ dollars. (Notice that the answer in (b) is slightly higher than that in (a).)

11 $\dfrac{1-m}{3m}$

12 $y = \dfrac{1}{81}, 9$

13 $\log_x(2) = \dfrac{1}{\log_2(x)}$, and $x = 2, 4$.

14 2

15 $\log(p) = 3a$ and $\log(q) = a$, so $\log_p q = \dfrac{1}{3}$

16 $a = \dfrac{1}{2}$, $b = \dfrac{1}{3}$. Hence, $(x, y) = (27, \dfrac{1}{3})$

17

(a) $x = \dfrac{1}{2}\log_2(10)$

(b) $y = -2$

18

(a) $y = 20$

(b) $x = \dfrac{1}{3}$

19

(a) $x = 4^u$

(b) $2 - u$

(c) $\dfrac{3}{2u}$

Topic 10

Sequence and Series

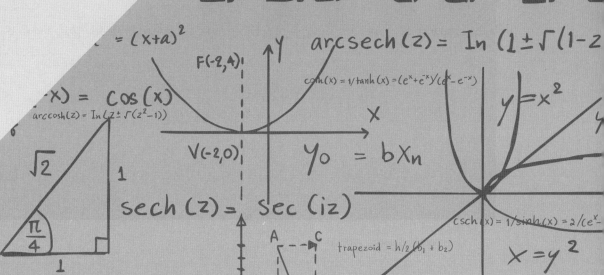

10.1 Sequence

A set of numbers, stated in a definite order, such that each number can be obtained from the previous number according to the unchanging rule or determined by a set of rules, is called a sequence. The set of whole numbers, for instance, is a sequence with consecutive difference of 1. Such rule is known as the recursive rule. Each number of the sequence is called a term.

- $3, 5, 7, 9, 11, \cdots$ is an arithmetic sequence.

- $1, 4, 9, 16, 25, \cdots$ is a sequence with perfect squares.

- $1, 2, 4, 8, 16, \cdots$ is a geometric sequence.

The expression u_n for the nth term of the sequence is useful since any specific term of the sequence can be obtained from it. The nth term of the sequence $1, 4, 9, 16, 25, \cdots$ is n^2. Thus, we could write $u_n = n^2$. For $n = 1$, $u_1 = 1^1 = 1$. For $n = 2$, $u_2 = 2^2 = 4$, and so on.

For some sequences, each term may be defined as a function of previous terms. One of the famous examples is the Fibonacci sequence. Leonardo Pisano Bigollo, also known as Fibonacci, was born in Europe around 1200's. One of his famous work, the Fibonacci sequence, starts with the first two terms of 0 and 1. Then, each subsequent term is the sum of the two terms before it. The first two terms are denoted by F_0 and F_1, so $F_0 = 0$ and $F_1 = 1$. The rest of the terms are naturally found by

$$F_n = F_{n-1} + F_{n-2}$$

so $F_2 = F_1 + F_0 = 1$, $F_3 = F_2 + F_1 = 2$, $F_4 = F_3 + F_2 = 3$, $F_4 = F_3 + F_2 = 5$, and so on. Since every term after the initial two terms is defined in terms of earlier terms in the sequence, we call this a recursive sequence.

1

(a) If the sequence $\{a_n\}$ satisfies $\dfrac{3}{4}, \dfrac{5}{8}, \dfrac{7}{16}, \dfrac{9}{32}, \cdots$, find the general formula for $\{a_n\}$.

(b) If the sequence $\{b_n\}$ satisfies $1, 1, 2, 3, 5, 8, 13, \cdots$, find the general rule for $\{b_n\}$ and find the next two terms.

10.2 Arithmetic Sequence and Series

An arithmetic progression (or arithmetic sequence) is a sequence of numbers where we find the next term in the sequence by adding the previous term by a number.

- Definition: $a_{n+1} - a_n = d$ where d is the common difference.

- $x = \dfrac{a+b}{2} \iff a, x, b$ form an arithmetic sequence, where x is known as arithmetic mean of a and b.

- Closed form: $a_n = a_1 + (n-1)d$ where d is the common difference.

- The sum of the first n terms:

$$S_n = \frac{n(a_1 + a_n)}{2} = \frac{n(2a_1 + (n-1)d)}{2}.$$

- $m + n = p + q \iff a_m + a_n = a_p + a_q$ for $m, n, p, q > 0$.

$\boxed{2}$ For the following arithmetic progressions $4, 9, 14, \cdots$, find the 20th term.

$\boxed{3}$ The 3rd term of an arithmetic progression is 13 and the 8th term is 33. Find the 27th term.

The fourth bullet point above, labeled as the sum of the terms of arithmetic sequence, is also called as "arithmetic series." This may be written as

$$S_n = x_1 + x_2 + x_3 + \cdots + x_n$$

whose formula for the arithmetic series can be deduced using some algebra.

$$S_n = x_1 + x_2 + x_3 + \cdots + x_{n-1} + x_n$$
$$S_n = x_n + x_{n-1} + x_{n-2} + \cdots + x_2 + x_1$$
$$\Downarrow$$
$$2S_n = (x_1 + x_n) + (x_2 + x_{n-1}) + (x_3 + x_{n-2}) + \cdots + (x_{n-1} + x_2) + (x_n + x_1)$$
$$2S_n = n(x_1 + x_n)$$
$$S_n = \frac{n(x_1 + x_n)}{2}$$

$\boxed{4}$ Bob earns an annual salary of $50,000$ for the first year of his employment. His annual salary increases by $1,200$ each year.

(a) Calculate the annual salary of the fifth year of his employment.

He remains in this employment for 10 years.

(b) Calculate the total salary he earns in this employment for 10 years.

10.3 Geometric Sequence and Series

A geometric progression (or geometric sequence) is a sequence of numbers where we find the next term in the sequence by multiplying the previous term by a number.

- Definition: $\dfrac{a_{n+1}}{a_n} = r$ where r is the common ratio.

- $x = \pm\sqrt{ab} \iff a, x, b$ form a geometric sequence, where x is the geometric mean of a and b.

- Closed form: $a_n = a_1 r^{n-1}$ where r is the common ratio.

- The sum of the first n terms:
$$S_n = \frac{a_1(1 - r^n)}{1 - r} \, (r \neq 1)$$

- If $r = 1$, then $S_n = na_1$.

- $m + n = p + q \iff a_m a_n = a_p a_q$ for $m, n, p, q > 0$.

⑤ For the geometric progression $\{3, 6, 12, \cdots, 3072\}$, find the number of terms.

⑥ If the following three expressions, $k+3$, $5k-3$, and $7k+3$ are three consecutive terms of a geometric progression, find the values of k.

An infinite geometric progression (or infinite geometric series) whose common ratio is between -1 and 1, not inclusive, has a convergent series value whose value is given by

$$S = \frac{u_1}{1-r}$$

where $|r| < 1$.

7 Find the sum of the following series $1 + \frac{1}{2} + \frac{1}{4} + \cdots$. (Assume that the infinite series is geometric.)

8 Find the sum of $3 + 1 + \frac{1}{3} + \cdots$. (Assume that the infinite series is geometric.)

Covered in previous chapter, the application of geometric sequence and series appears again in this chapter. Exponential growth or decay can be written in geometric sequences. For instance, the population of bacteria doubles every hour. Or, a bank account gives 5 percent annual interest rate every year. Exponential growth occurs when the amount at each stage is multiplied by a constant greater than 1. In the case of bacteria growth, the constant is 2; the bank account has the constant of 1.05.

9 The population of white birches in a forest has been decreasing by 2 percent every year. The population at the beginning of 2007 was estimated to be 10,000. If P is the population of trees t years after 2007, which of the following equations is the correct population of trees over time t?

(A) $P(t) = 10,000(0.02)^t$
(B) $P(t) = 10,000(0.98)^t$
(C) $P(t) = 10,000(0.98)^{4t}$
(D) $P(t) = 10,000(0.98)^{t/4}$

10 The bank account has 3 percent annual interest rate compounded every year. If the initial deposit is 1,000 dollars, what is the amount in the bank account after 3 years? Set up the equation and compute the exact value with a calculator.

10.4 Arithmetic Mean and Geometric Mean

In the previous sections, we learned about arithmetic and geometric sequence. Let's see how they earned these names. First, if three numbers are in arithmetic progressions, i.e., $\{a,b,c\}$, then b is called the arithmetic mean of a and c, which means that each term is the arithmetic mean, or average of its nearest neighbors.

$$b - a = c - b$$
$$2b = a + c$$
$$b = \frac{a+c}{2}$$

On the other hand, if three positive numbers are in geometric sequence, i.e., $\{a,b,c\}$, then b is called the geometric mean of a and c. By the definition of geometric progression, we know that

$$\frac{b}{a} = \frac{c}{b}$$
$$b^2 = ac$$
$$b = \sqrt{ac}$$

Not only are we able to find arithmetic mean or geometric mean, but also we can insert a few terms as arithmetic mean or geometric mean between any two numbers. These two mean values may be equal or not. If two terms are equal, then the arithmetic or geometric mean share same values. Otherwise, the arithmetic mean is likely to be larger than the geometric mean, which is also known as AM-GM inequality.

$\boxed{11}$ Find the arithmetic mean and geometric mean of 2 and 18.

$\boxed{12}$ The first three terms of an arithmetic sequence are $2x + 3$, $5x - 2$ and $10x - 15$. Find the value of x.

10.5 Difference Sequence

If $x_n = an + b$ where a and b are constants, then $\{x_n\}$ is an arithmetic sequence. What will happen if we extend the expression into higher powers?

A quadratic sequence has the general term $x_n = an^2 + bn + c$ and a cubic sequence has a general term $x_n = an^3 + bn^2 + cn + d$. In order to find the formula for these sequences, we may use two types of techniques.

- Substitution

- Difference Method

There is nothing to learn from substitution, since we did this since Algebra 1. On the other hand, the difference method is all about finding the consecutive differences until the difference stays constant.

- Linear Sequence $x_n = an + b$.

n	1	2	3
x_n	$a+b$	$2a+b$	$3a+b$
$\triangle_1$		$1! \times a$	$1! \times a$

- Quadratic Sequence $x_n = an^2 + bn + c$

n	1	2	3	4
x_n	$a+b+c$	$4a+2b+c$	$9a+3b+c$	$16a+4b+c$
$\triangle_1$		$3a+b$	$5a+b$	$7a+b$
$\triangle_2$			$2! \times a$	$2! \times a$

where $\triangle_i$ indicates the consecutive difference in the ith row.

13 Use the difference method to find $x_n = an^2 + bn + c$ for the following sequences.

(a) $\{0, 2, 6, 12, \cdots\}$.

(b) $\{2, 5, 10, 17, \cdots\}$.

14 Use the difference method to find the general term a_n of

(a) $\{5, 9, 13, 17, 21, \cdots\}$.

(b) $\{6, 12, 20, 30, 42, \cdots\}$.

(c) $\{2, 7, 18, 38, 70, 117, \cdots\}$.

10.6 The Σ Notation

The sum of the terms of a sequence is called a series. In Algebra 2, we learn mostly about arithmetic series (the sum of arithmetic sequences) and geometric series (the sum of geometric sequences). Before talking about these two series, we will learn about the summation signs.

The large Greek symbol, $\sum$, is a summation sign that represents series. Let's look at the following example.

$$\sum_{i=1}^{4} i^2 = 1^2 + 2^2 + 3^2 + 4^2$$

The expression $i = 1$ specifies where we begin with for the summation. The top denotes where i ends. The dummy variable i in the expression above usually increases by 1, and the values of expression next to $\sum$ are added for all the different values from $i = 1$ to $i = 4$. The following formula list is useful when dealing with series questions, and they can be proven by the principle of induction.

Series Formula

$\checkmark$ $\displaystyle\sum_{k=1}^{n} k = \frac{n(1+n)}{2} = 1 + 2 + 3 + \cdots + (n-1) + n.$

$\checkmark$ $\displaystyle\sum_{k=1}^{n} k^2 = \frac{n(n+1)(2n+1)}{6} = 1^2 + 2^2 + 3^2 + \cdots + (n-1)^2 + n^2.$

$\checkmark$ $\displaystyle\sum_{k=1}^{n} k^3 = \left(\frac{n(n+1)}{2}\right)^2 = 1^3 + 2^3 + 3^3 + \cdots + (n-1)^3 + n^3.$

---

15 Without using a calculator, evaluate the following series.

(a) $\displaystyle\sum_{k=1}^{20} (3k+4)$

(b) $\displaystyle\sum_{k=1}^{15} (2k^2+4)$

16 Without using a calculator, evaluate the following series.

(a) $\displaystyle\sum_{k=11}^{25} k^3$

(b) $\displaystyle\sum_{k=6}^{18} (k+1)(k+2)$

(c) $\displaystyle\sum_{k=1}^{8} 2^{-k}$

17 If $S_n = n^2$, find a_n, the nth term of the sequence.

1 Consider the following sequence $\{57, 55, 53, 51, \cdots, 5, 3, 1\}$.

(a) Find the number of terms of the sequence.

(b) Find the sum of the sequence.

2 The first term of an arithmetic sequence is 4 and the sum of the first two terms is 11.

(a) Write down the second term of the sequence.

(b) Find the common difference of the sequence.

(c) If the nth term is the first term greater than $1,000$, find the value of n.

3 Calculate the sum of natural numbers from 1 to 200.

4 The fifth term of an arithmetic sequence is 20 and the twelfth term is 41.

(a) Find the hundredth term.

(b) Find the sum of the first 100 terms.

5 A geometric sequence $\{x_n\}$ satisfies $x_7 = 108$ and $x_8 = 36$.

(a) Find the common ratio of the sequence.

(b) Find x_1.

(c) If the sum of the first n terms is $118,096$, find the value of n. If there is no such n, then write down "it does not exist."

6 Consider a geometric sequence, $\{16, 8, a, 2, b, \cdots\}$.

(a) Find the common ratio.

(b) Find the values of a and b.

(c) Find the sum of the first 6 terms.

7 Given a geometric sequence $\{8, x, 2, \cdots\}$ for which the common ratio is $\dfrac{1}{2}$,

(a) Find the value of x.

(b) Find the value of the 9th term.

(c) Find the sum of the first 10 terms.

8 Bob starts his first job. His annual salary in the first year is $26,000 and his salary increases 3% every year.

(a) Compute how much he will earn in his third year of working.

Bob spends $24,800 of his earnings. For the next 10 years, inflation will cause his living expenses to rise by 5% per year.

(b) Compute the number of years it will be when Bob spends more than he earns.

9 Use the difference method to find the general term x_n of $-1, 2, 7, 14, 23, \cdots$.

10 Without using a calculator, evaluate

(a) $\displaystyle\sum_{j=1}^{10} (k^2 + 3k + 1)$

(b) $\displaystyle\sum_{k=3}^{10} (3k - 5)$

(c) $\displaystyle\sum_{n=1}^{20} (4m)$

1

(a) 29   (b) 841

2

(a) $a_2 = 7$ (b) 3   (c) 334

3 20100

4

(a) 305   (b) 15650

5

(a) $\dfrac{1}{3}$

(b) $4 \times 3^9 = x_1$

(c) Solving for $118096 = \dfrac{4 \times 3^9 (1 - 1/3^n)}{1 - 1/3}$ produces $n = 10$.

6

(a) $\dfrac{1}{2}$ (b) $a = 4, b = 1$ (c) $31\dfrac{1}{2}$

7

(a) 4   (b) $\dfrac{1}{32}$ (c) $\dfrac{1023}{64}$

8

(a) $26000(1.03)^2 \approx 27583.4$, in dollars.

(b) Solving for $24800(1.05)^{n-1} > 26000(1.03)^{n-1}$, $n > 1 + \dfrac{\log(26000/24800)}{\log(1.05/1.03)} \approx 3.46$. Hence, Bob spends more than he earns roughly after three and half years.

9 $x_n = n^2 - 2$

10

(a) $10(k^2 + 3k + 1)$ (b) 116   (c) $80m$

Topic 11

Counting : Permutation

and Combination

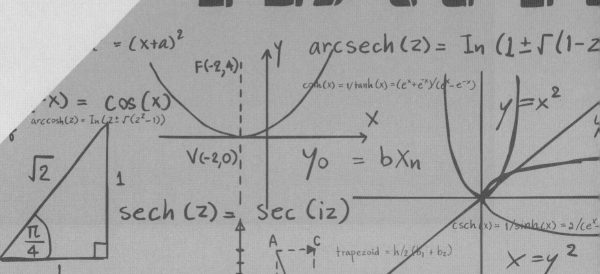

11.1 From a to b

The basic counting starts with 1 and ends at the last integer. What if I ask you to find the number of integers between 45 and 317, exclusive? Then, we can count one by one, starting from $46, 47, \ldots, 316$. Is there a smarter way to compute this? The answer is YES. In fact, if you subtract the two given numbers and add by one, we get the number of integers between the given ones. However, it is better for us to learn 1-to-1 correspondence concept using this example, illustrated as in the following arrays.

$$46, 47, \ldots, 316$$
$$1, 2, \ldots, 271$$

The first array of numbers and the second array of numbers have a 1-1 correspondence. Hence, computing the numbers from 1 to 271 is equivalent to counting numbers from 46 to 316. Notice that $46 - 1 = 42 - 2 = \cdots = 316 - 271 = 45$.

1 Find the number of integers between 56 and 174 inclusive.

2 Find the number of even integers in $2, 4, 6, \cdots, 432$.

11.2 Multiplication Principle

We multiply if

- the occurrences are *repetitive*. For instance, if we flip two coins, all possible results are HH, HT, TH, TT. Notice that, given any result of the first coin, either H or T, the second coin always has two results H or T. In other words, ___T or ___H will occur for any letter that goes inside the first underline. Hence, we multiply 2×2 where the first 2 represents all possible outcomes of the first coin, while the second 2 shows all possible outcomes of the second coin.

- the occurrences are *successive*. Come back to our coin flipping cases. If we imagine flipping coins separately, there must be the first and second coin to be flipped. What is the number of possible outcomes of the first coin? It is 2. Are we done? No! We must flip the last coin, too! Flip the last coin, where the number of possible outcomes of the second coin is also 2. When action is ongoing, we multiply the number of outcomes.

$\boxed{3}$

(a) How many four-letter words are there? (There are 26 alphabets to begin with, and assume that four-letter words in this question do not have to make sense.)

(b) A store sells 4 distinct pants, 3 different shirts, and 2 different pairs of shoes. How many different apparel styles are possible? (Assume that one may choose one pant, one shirt, and one pair of shoes.)

(c) A spinner with four equal sections is spun three times. How many outcomes are possible? (Assume that four sections are all distinguishable.)

When there are *restrictions*, we count the restrictions first. This is really convenient because we can multiply in any order we consider proper for the given situation.

4

(a) How many even 5-digit numbers are possible if there is no repetition of digits?

(b) How many four-letter words are there with a vowel at the end and consonants in the middle two places?

(c) Suppose there are five digits $\{1,2,3,4,8\}$. How many four-digit numbers are greater than 4200 if repetition is not allowed?

5

(a) In how many ways can Jimmy give Skittles to three kids? (Plentiful skittles come in red, orange, yellow, green, and purple, and assume that any kid gets one skittle.)

(b) If the final score in a soccer match is 2 : 0, how many different scores are possible at the end of half-time?

6

(a) How many odd numbers with third digit 5 are there between 20000 and 69999, inclusive?

(b) Consider the letters I,R,O,N,M,A,N. How many different four-letter words can be formed if repetition is not allowed? (Assume that these all of these two N's can be used.)

We can connect Number Theory to counting, i.e., the number of divisors. Using the principle of multiplication, we can simply calculate the number of divisors by counting. In Number Theory, the number of divisors of a positive integer n is usually denoted by $d(n)$.

(a) How many positive divisors does 24 have?

(b) How many positive divisors does $1,000,000$ have?

(c) If $n = p^3 q^4$ for distinct primes p and q, find the number of positive divisors of n.

Simple multiplication only works when all contributors are completely independent. This fails for a problem like:

In how many ways can David give five differently-colored Skittles to three kids?

The answer is not 5^3. The reason that independence breaks down here is that there are a limited number of Skittles. If David gives a red skittle to the first kid, the option no longer exists of giving a red to any of the others.

In such a situation, we can still use multiplication, but incorporating the restriction that there is only one of each color. So we line the kids up (in our head), and start giving out candy. There are 5 ways to give the first kid a piece, then only 4 for the second, since one is gone (does not matter which), then 3 for the third.

(a) In how many ways can a five-letter word be written using only the first half of the alphabet with no repetitions such that the third and fifth letters are vowels and the first a consonant?

(b) Consider the digits $0, 1, 2, 3, 4, 5, 6$. How many 4-digit numbers can be formed if repetition is not allowed and the number must be divisible by 5? (Assume that 4-digit numbers are positive.)

11.3 Permutation : Counting Successive Events

Permutation is the number of arrangements of k seats with n different people, denoted by $_nP_k$ and computed by

$$n(n-1)(n-2)\cdots(n-k+1), \quad \text{or} \quad \frac{n!}{(n-k)!}.$$

Factorial is defined by $n! = n(n-1)(n-2)\cdots 2\cdot 1$. Basic math problems may involve the definition of factorials, whereas intermediate ones use permutation as a tool.

9 Evaluate the following expressions without using a calculator.

(a) $_6P_4$

(b) $_{10}P_2$

(c) $_4P_4$

10 Solve for n.

(a) $2 \times {_nP_2} = 3 \times {_{n-1}P_2}$

(b) $_{n+1}P_2 = 50 \times {_nP_1}$

11 Find the number of permutations of the letters in the word PANTHER if

(a) the vowels must be together.

(b) the letters P and T must not be adjacent.

12 Five people, Abraham, Blair, Charles, Daniel, and Elliot, are seated in a row of chairs. In how many ways can they be seated if

(a) Abraham and Blair must be seated together?

(b) Blair, Charles, and Elliot must be seated together?

Let's extend our argument. Permutation is an arrangement of objected in a line. If we arrange objects *in a line*, then we have no worries.

$$A - -B - -C \quad B - -C - -A \quad C - -A - -B$$

are considered distinct. But what about things laid out in a circle? How about the following figures?

$$\text{A} \qquad \text{C} \qquad \text{B}$$

$$\text{B} \quad \text{C} \qquad \text{A} \quad \text{B} \qquad \text{C} \quad \text{A}$$

Why are these three apparently different arrangements considered to be the same? Consider what person A sees in each case: B on the right, C on the left. To A, the arrangement looks the same in all cases. If you consider what B and C see, you will see that the three cases are equivalent to them as well.

On the other hand, reflections do matter, since after a reflection, A sees B on her left! Thus

$$\text{A}$$

$$\text{C} \quad \text{B}$$

is different from the previous three. There are only two distinct circular arrangements of 3 objects. If we count objects in a circle as we do objects in a line, we decide there are $n!$ arrangements. However, as shown above, these arrangements are not all distinct. Each distinct arrangement is counted n times, once for each rotation of the objects. In our case,

$$A, B, C$$
$$B, C, A$$
$$C, A, B$$

They should be considered as over-counted. Hence, we must divide the number of arrangements by n, producing $n!/n = (n-1)!$ distinct arrangements. In the following four problems, assume that one configuration is considered equivalent to another one if rotated to be indistinct.

13 In how many ways can five students be seated around a circular table with equally spaced seats?

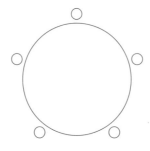

14 In how many ways can six people, A, B, C, D, E, and F, be seated in a circular way if A and B must sit next to each other?

15 In how many ways can three couples be seated around a circular table if each couple must sit next to one another?

16 In how many ways can three adults and three children be seated around a circular table if an adult and a child must alternate?

11.4 Principle of Addition

In general, we add the number of counts if

- we have to use *caseworks*.

- the occurrences are *not repetitive*.

If certain actions or events cannot be performed simultaneously, then we simply add the number of ways altogether. Also, if the structure of counting turns different, we also add.

17 In how many ways can Janet give one Skittle each to two children if she has 3 different red, 4 different brown, and 5 different tan ones and the two children insist upon having Skittles of different colors? (Assume that each child must take one skittle.)

18 How many odd numbers with middle digit 5 and no digit repeated are there between 20000 and 69999, inclusive?

Many counting problems can be subtly changed by the distinguishability of the objects counted. Consider the problem:

In how many ways can two objects be put into two boxes?

If the objects are distinguishable, there are four different ways: both in box 1, object A in box 1 and B in 2, B in 1 and A in 2, and both in 2. But if the objects are indistinguishable (meaning we can't tell the difference between them), the two middle choices are the same, and there are only three cases!

We can make it worse by making the boxes themselves indistinguishable. In this case, "both in one box" is only one choice, since we don't care which box we are talking about. Similarly, "one in one box and one in the other" is a single choice, even if the objects are themselves distinguishable. This is called the partition of natural numbers. Try to think about 1's. They are all indistinguishable between one another, but if there are two 1's, they make 2.

$$2 = 2 + 0$$
$$= 1 + 1$$

As one can check the partition of natural numbers tells us the number of putting indistinguishable objects into indistinguishable boxes.

19 In how many ways can three different babies be put in two different playpens? Or, in two identical-looking ones?

20 In how many ways can three identical-looking rattles be given to two different babies? (Apply the partition of natural numbers.)

[21] A semi-final for soccer championship is won when a team wins three of four matches. How many possible outcomes for the team to go up to the finals? (Assume there is no draw, and the semifinal result is determined as long as whichever team wins three matches first out of a total of four matches.)

[22] Suppose there is a panel of four light switches in a row. How many settings are there if two consecutive switches cannot both be turned off?

11.5 Combination : Another Tool for Overcounts

Combination is the number of selection of r objects from n different objects. The order in this counting does not matter. In fact, we overcount using permutation and get rid of the overcounts by division. Sometimes, we overcount in order to solve harder questions.

$$_nC_r = \binom{n}{r} = \frac{n!}{(n-r)!r!} = \frac{_nP_r}{r!}$$

where $n \geq r \geq 0$.

23 Simplify the following combinations.

(a) $_5C_2$ (b) $_5C_5$ (c) $_5C_0$

24 Solve for n where $_nC_2 = 55$.

25 In how many ways can Cleopatra choose 2 different colored Skittles out of 5 differently-colored Skittles to eat?

As one can see from **25**,

$$_nC_r = \frac{_nP_r}{r!}$$

This tells us the usage of "division." The purpose of division is to eliminate the overcounts. This result is the number of combinations of k objects from a set of n objects. It is denoted by $\binom{n}{k}$, or sometimes by $_nC_k$.

26 In how many distinguishably different ways can a pair of indistinguishable dice come up?

Also, combination satisfies a peculiar property known as

$$_nC_r = {_n}C_{n-r}$$

The short proof of it can be given by

$$_nC_{n-r} = \frac{n!}{(n-(n-r))!(n-r)!}$$
$$= \frac{n!}{r!(n-r)!}$$
$$= {_n}C_r$$

27 Without using a calculator, evaluate the following combinations using the property stated above.

(a) $_{100}C_{99}$

(b) $_{n+1}C_{n-1}$

28 A set of points is chosen on the circumference of a circle so that the number of different triangles with all three vertices among the points is equal to the number of pentagons with all five vertices in the set. How many points are there?[1]

---

[1]There are infinitely many points on a circle, but we specifically denote some number of points to form triangles and pentagons in this problem.

11.6 Permutation vs. Combination

Don't confuse combinations with permutations! The order of objects in permutation matters, whereas that in combination does not. So if you want an ice-cream cup, but don't care about the order of the flavors, the number of cups you can make is a problem in combinations. On the other hand, if you want an ice-cream cone that distinguishes which scoop is on top and which is on bottom, the number of cones you can make is a problem in permutations.

The overcounting concept is still useful. For example, suppose we want to find in how many ways the word NAGUINI can be rearranged. We might immediately think the answer is 7!, since there are 7 letters in the word. The problem is that we are over-counting. We counted too much by considering $I_1 \neq I_2$ and $N_1 \neq N_2$, though they are indistinguishable.

Among the 7! arrangements, each I and N has been written 2! times by ordering them in 2! ways. Dividing out by all these overcounts, since we only want one copy of each arrangement in the end, the final answer is $\frac{7!}{2!2!}$.

29 In how many ways can the word *RAMANUJAN* be rearranged? How about *MINIMIZATION*?

Combination is exactly the same concept of filling the positions in a line with indistiguishables. In fact, if you feel comfortable with counting problems, a good exercise is to do problems in more than one way. This helps you stay flexible and not get locked into one mode of solving counting problems.

30 Given $ABCDEFAAB$, find the the number of possible arrangements.

11.7 Binomial Theorem

$$(x+y)^n = \binom{n}{0}x^n + \binom{n}{1}x^{n-1}y + \binom{n}{2}x^{n-2}y^2 +$$
$$\cdots + \binom{n}{n-1}xy^{n-1} + \binom{n}{n}y^n$$

Notice that we put the $\binom{n}{0}$ and $\binom{n}{n}$ in the first and last terms to make the pattern complete. This expansion is known as the Binomial Theorem, and it is the quickest way to evaluate powers of binomial expressions.

31 Write the first three terms in the expansion of the following binomial expressions.

(a) $(2x+3y)^5$

(b) $(x+\dfrac{1}{x})^4$

(c) $(3x+y)^{10}$

32 Find the constant term of the expansion of $\left(x^2 - \dfrac{2}{x}\right)^6$.

Binomial Theorem may be applied with a concept called "generating function," which changes counting problems into finding exponents in polynomial expansion. Let's suppose there are two people with a dollar each and a third person with two dollars. Suppose that two people with a dollar each may choose to pay nothing or one full dollar. Likewise, suppose that the last person with two dollars may choose to pay nothing, a dollar, or two dollars, assuming that all three have dollar bills instead of coins. Have a look at the following example of polynomial expansion.

$$(x^0 + x^1)(x^0 + x^1)(x^0 + x^1 + x^2) = (1+x)(1+x)(1+x+x^2)$$
$$= (x^2 + 2x + 1)(x^2 + x + 1)$$
$$= x^4 + 3x^3 + 4x^2 + 3x + 1$$

Let's investigate what this has to do with our counting. There is no meaning to a variable x, but exponents matter.

- The coefficient of x^4 is 1. This means, in our context, there is only one way of paying 4 dollars in total.

- The coefficient of x^3 is 3. This means, in our context, there are three ways of paying 3 dollars in total. If we let the amount paid by 1st, 2nd, and 3rd person as (p_1, p_2, p_3), then $(p_1, p_2, p_3) = (1, 0, 2), (0, 1, 2), (1, 1, 1)$.

- The coefficient of x^2 is 4. This means, in our context, there are four ways of paying 2 dollars in total. Borrowing the same letters from the previous bullet point, we get $(p_1, p_2, p_3) = (1, 1, 0), (1, 0, 1), (0, 1, 1), (0, 0, 2)$.

- The coefficient of x is 3. This means, in our context, there are three ways of paying 1 dollar in total. We get $(p_1, p_2, p_3) = (1, 0, 0), (0, 1, 0), (0, 0, 1)$.

- The constant term is 1. This means, in our context, there is one way of paying nothing in total. We get $(p_1, p_2, p_3) = (0, 0, 0)$.

$\boxed{33}$ Ten people with one dollar each and one person with three dollars get together to buy an eight-dollar pizza. In how many ways can they do it?

1 What is the total number of digits used when the first 1234 positive even integers are written? (If first 2002 positive even integers are written, then there are 7456 integers, which can be found by using 1-to-1 correspondence.)

2 Bob counts up from 1 to 9, and then immediately counts down again to 1, and then back up to 9, and so on, alternately counting up and down

$$(1,2,3,4,5,6,7,8,9,8,7,6,5,4,3,2,1,2,3,4,\ldots).$$

What is the 1000th integer in his list?

3 How many different rectangles with sides parallel to the grid can be formed by connecting four of the dots in a 4×4 square array of dots, as in the figure below?

$$\begin{matrix} \cdot & \cdot & \cdot & \cdot \\ \cdot & \cdot & \cdot & \cdot \\ \cdot & \cdot & \cdot & \cdot \\ \cdot & \cdot & \cdot & \cdot \end{matrix}$$

(Two rectangles are different if they do not share all four vertices.)

4 Find the largest integer n for which 2^n evenly divides $20!$.

5 The number 64 has the property that it is divisible by its units digit. How many whole numbers between 10 and 50 have this property?

6 If $1 \le x \le 10$ and $1 \le y \le 36$, for how many ordered pairs of integers (x, y) is $\sqrt{x + \sqrt{y}}$ an integer?

7 In how many ways can 8 people be seated in a row of chairs if at least two of three people "Abraham, Paul, and Richard" refuse to sit in adjacent seats?

8 The sequence 2, 3, 5, 6, 7, 10, 11, ... contains all the positive integers from least to greatest that are neither squares nor cubes. What is the 20th term of the sequence?

9 The Goblin's language consists of 3 words, "Argh," "Glur," and "Hark." In a sentence, "Argh" cannot come directly before "Glur"; all other sentences are grammatically correct (including sentences with repeated words). How many valid 3-word sentences are there in this language?

1 4384

2 8

3 36

4 18

5 17

6 10

7 14400

8 26

9 21

Algebra 2 개념학습과 더불어 도움이 되는 "harimath.com"

> Scott Young이 저술한 〈울트라러닝〉을 읽고, 깊은 울림을 경험한 저자가 웹 언어 코딩을 학습하여, 2022년 8월부터 9월 사이에 몰입하여 제작한 웹사이트입니다.
>
> 해당 웹사이트는 두 가지 목적을 가지고 있습니다.
>
> 첫 번째, Essential Math Series 커리큘럼으로 집필된 교재의 오타 및 오류 등은 발견되는 대로, harimath.com에서 errata 탭에서 확인이 가능합니다. 수시로 업데이트가 되므로, 자주 확인하길 바랍니다.
>
> 두 번째, 저자가 수시로 제작한 수학 경시대회 문제들을 업로드 해두었습니다. 경시대회를 준비코자 하는 학생들과 학교 시험에서 고난이도 문제로 출제되는 내용에 도움을 주고자, 다양한 문제들을 계속 업데이트 중이며, 각 문제들에 대한 answer key 제공을 하나, full solution은 제공하지 않으므로 인출 효과(Retrieval Method)를 극대화합니다.
>
> Algebra 2까지 끝난 학생들 혹은 수학 경시대회를 준비하는 학생들에게 도움이 되는 웹사이트이므로, 학창시절 도움이 되는 resource로 잘 활용하길 바랍니다!

Topic 12

Probability

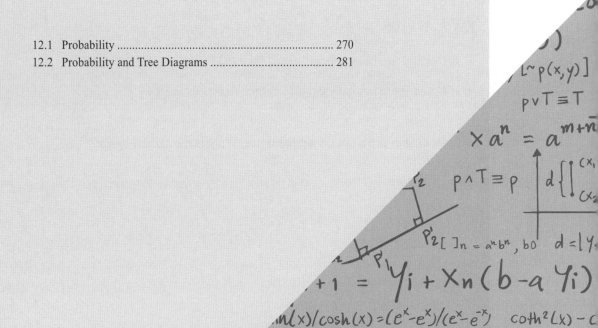

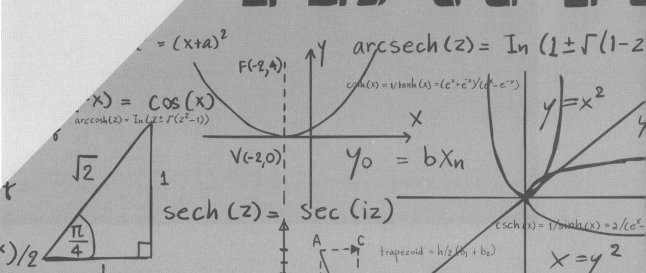

12.1 Probability

A probability of an event U to occur is between 0 and 1. If $P(U) = 0$, then U will never happen. On the other hand, if $P(U) = 1$, then U must happen. In order to solve probability questions, we use the following five laws of probability or solve the problems by drawing a Venn Diagram and some intuition.

- $P(U') = P(\text{not } U) = 1 - P(U)$

- $P(U \cup V) + P(U \cap V) = P(U \text{ or } V) + P(U \text{ and } V) = P(U) + P(V)$

- $P(U|V) = P(U \text{ given } V) = \dfrac{P(U \cap V)}{p(V)}$

- If U and V are independent events, then $P(U \cap V) = P(U) \times P(V)$.

- If U and V are mutually exclusive, then $P(U \cap V) = 0$.

$\boxed{1}$ Bob is planning to flip a fair coin three times. What is the sample space? [1]

$\boxed{2}$ Mr. Know-it-all rolls two fair dice. What is the probability that the sum will be 6 or less?

---

[1] Sample space consists of all possible events.

$P(A|B)$ is the probability of A given that B has already occurred. It can be found by

$$P(A|B) = \frac{P(A \cap B)}{P(B)} = \frac{\text{what we want}}{\text{what is given}}$$

In other words, it is the proportion of $A \cap B$ from B. As one can see from the Venn Diagram, it is the proportion

$$\frac{\text{Darker}}{\text{Darker} + \text{Lighter}}$$

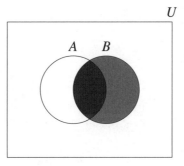

3 If $P(A) = \dfrac{1}{2}$ and $P(B) = \dfrac{3}{8}$ and $P(A \cap B) = \dfrac{1}{8}$, then find

(a) $P(\text{not } A)$ (b) $P(A \text{ or } B)$ (c) $P(A \text{ given } B)$

4 If $P(A) = \dfrac{1}{3}$ and $P(B) = \dfrac{5}{18}$ and $P(A \cup B) = \dfrac{5}{9}$, then find

(a) $P(A \cap B)$ (b) $P(A|B)$

(c) Find whether A and B are independent.

(d) Find whether A and B are mutually exclusive.

$\boxed{5}$ If $P(A \cup B) = \dfrac{4}{5}$ and $P(A \cap B) = \dfrac{1}{5}$ and $P(A|B) = \dfrac{3}{5}$, then find

(a) $P(A)$ (b) $P(B)$

(c) $P(B|A)$ (d) $P(B|A')$

(e) $P(B'|A')$ (f) $P(A \cap B|A \cup B)$

6 Two fair dice are rolled. Fill the empty blanks in the score table.

	1	2	3	4	5	6
1						
2						
3						
4						
5						
6						

Hence, find

(a) P(the sum of scores is 8 or more)

(b) P(the sum of scores is divisible by 3)

(c) P(the sum of scores exceeds 9)

(d) P(the sum of scores is less than 6)

7 Suppose there are six oranges, four apples, and three bananas in a box. Choose four without replacement.

(a) P(no bananas)

(b) P(all oranges)

(c) P(one apple)

(d) P(all same fruits)

When we are dealing with probability questions, we encounter "at least" phrases. There are more such similar to this one as followed.

- $P(\text{at least } 5) = P(5,6,7,...) = 1 - P(0,1,2,3,4)$
- $P(\text{more than } 2) = P(3,4,5,...) = 1 - P(0,1,2)$
- $P(\text{no more than } 3) = P(0,1,2,3) = 1 - P(4,5,6,...)$
- $P(\text{less than } 4) = P(0,1,2,3) = 1 - P(4,5,6,...)$

8 A bag contains 4 red, 4 white, and 2 blue balls. Three are drawn without replacement. Find

(a) P(at least one red)

(b) P(no more than one red)

(c) P(less than one red)

(d) P(more than 2 white)

(e) P(at least 2 white)

9 When 4 cards are drawn from a deck of 52 cards without replacement, find

(a) P(at least one King)

(b) P(more than 3 spades)

(c) P(less than 2 diamonds)

(d) P(all with same face value)

10 The weather forecast shows a 55% chance of rain. If it rains, nobody goes to the beach. If it does not rain, then there is a 70% chance of going to the beach. In this case, what is the probability of anyone going to the beach?

Sometimes, we use constant probability when the probability stays unchanged for each event, unlike a deck of cards without replacement or balls drawn from a bag without replacement. This is known as binomial distribution.

11 The probability that Jacob, a PFC in the Army, hits the target in the range always equals to $\frac{2}{5}$. He fires six rounds at a shooting. Find

(a) P(exactly two hits)   (b) P(hit less than twice)   (c) P(hit at least twice)

12 In a hypothetical world with perfect genetic control, if $P(\text{boy born}) = \frac{2}{5}$, find

(a) P(3 boys in a family of 4 children)

(b) P(less than 3 girls in a family of 5 children)

13 A bag contains 4 red, 4 white, and 2 blue balls. Three balls are drawn from the bag without replacement. Find

(a) P(all red)   (b) P(2 reds)

(c) P(1 red)   (d) P(0 red)

14 Suppose three fair dice are rolled at the same time. Find

(a) P(all same face value)

(b) P(all the scores are different)

15 Four cards are drawn, without replacement, from a deck of 52 cards. Find

(a) P(all same suit)   (b) P(all different suits)

16 If five cards are drawn from a deck of 52 cards without replacement. Find

(a) P(3 Kings and 2 Queens in that order)

(b) P(3 Kings and 2 Queens in any order)

(c) P(2 Kings, 2 Queens, and a Jack)

17 I have 3 a's, 4 b's and 2 c's. I choose 3 out of these without replacement. Find

(a) P(all b's) (b) P(1 a) (c) P(2 c's)

REMARKS on Multiplication

Remember that when we count the probabilities, we ALWAYS look at whether we have the same kinds or not.

- If we have SAME kinds, then we do NOT multiply by the number of possible arrangements of the items.

- If we have DIFFERENT kinds, then we DO multiply by the number of possible arrangements of the items.

Example

In two-coin flip, compute the following probabilities.

(a) P(two heads) (b) P(one head and one tail)

Solution

(a) $\frac{1}{2} \times \frac{1}{2} \times 1$ (b) $\frac{1}{2} \times \frac{1}{2} \times 2$

12.2 Probability and Tree Diagrams

Remember that *or* means addition and *and* means multiplication.

18 There is a bag containing 4 red and 5 white balls. Two balls are drawn without replacement. Find the probability of

(a) 2 red balls

(b) a white, then red ball

(c) one of each color

(d) at least one red ball

19 If the probability that a student at You-know-what school is right-handed is 8/10, and the probability that a student wears glasses is 4/13. If two events are independent of one another, find the probability that a student at You-know-what school is left-handed and wearing glasses.

20 In you-know-what school, the probability that a student has brown eyes is 0.4 and, independently, the probability that a student is left-handed is 0.08. Calculate the probability that a student in you-know-what school

(a) does not have brown eyes and is not left-handed

(b) has brown eyes and is not left-handed.

21 At a party, the probability that Helen will win the best-wit prize is 0.05. The probability that she will win the best-dresser prize is 0.15.[2] Draw a tree diagram and hence find the probability

(a) that Helen loses both.

(b) that Helen wins only one of the two prizes.

---

[2]Assume that the two events are independent. In other words, the probability of Helen winning the best-wit prize does not interfere with the probability of her winning the best-dresser prize.

Bayesian Theorem

$$P(A|B) = \frac{P(A \cap B)}{P(B)} = \frac{P(B|A)P(A)}{P(B)} = \frac{P(B|A)P(A)}{P(B|A)P(A) + P(B|A')P(A')}$$

22 Suppose there are two identical looking bags.

4R, 6G	7R, 3G
Bag 1	Bag 2

A bag is chosen at random and a ball is selected at random.

(a) What is the probability that the ball is red?

(b) Given that the ball chosen is red, what is the chance that it was taken from bag 1?

1 If events M and N are such that $P(M) = 0.3$ and $P(N) = 0.4$,

(a) find the value of $P(M \cup N)$ when M and N are mutually exclusive.

(b) find the value of $P(M \cup N)$ when M and N are independent.

(c) Given that $P(M \cup N) = 0.6$, find $P(M \mid N)$.

2 In a class of 20 students, 12 students study Mathematics, 15 students study Chemistry and 2 students study neither subjects.

(a) Illustrate this information on a Venn Diagram.

(b) Find the probability that a randomly selected student from this class is studying both subjects.

(c) Given that a randomly selected student studies Chemistry, find the probability that this student also studies Mathematics.

3 If $P(X) = \dfrac{1}{6}$, $P(Y) = \dfrac{1}{3}$, and $P(X \cup Y) = \dfrac{5}{12}$, find $P(X^c \mid Y^c)$.

4 Given two events A and B, if $P(A) = 0.6$, $P(A \cup B) = 0.9$, and $P(A \mid B) = 0.6$, find $P(B)$.

5 If today is rainy, the chance that the next day is rainy is $\frac{2}{3}$. However, if today is sunny, the probability that the next day is rainy is $\frac{1}{4}$. Assume today is rainy. Represent this as a tree diagram for the next three days. Hence, find the probability that

(a) the next three days are all rainy.

(b) two of the next three days are sunny.

(c) one of the next three days is sunny.

(d) the third day is rainy.

(e) if the third day is rainy, the first day is also rainy.

6 A fair die is rolled. Given that it is less than 4, what is the probability that it is a score of 3?

7 When four cards are drawn from a deck of 52 cards, given that I drew at least 2 Kings, what is the probability that I drew 4 kings?

8 Four boys and three girls stay in a room. Two are called out at random.

(a) What is the probability that both are boys?

(b) Given that they are of the same gender, what is the probability that both are boys?

(c) Given that at least one girl is called out, what is the probability that two girls are called out?

9 Bob goes to school by bus every day. If it does not rain, the probability that the bus is late is $\dfrac{3}{20}$. If it rains, the probability that the bus is late is $\dfrac{7}{20}$. The probability that it rains in any day is $\dfrac{9}{20}$. On Monday the bus is late. Find the probability that it is not raining on Monday.

1

(a) 0.7   (b) 0.58   (c) 1/4

2

(a)

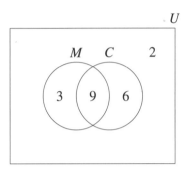

(b) $\dfrac{9}{20}$

(c) $\dfrac{3}{5}$

3 $\dfrac{7}{8}$

4 $\dfrac{3}{4}$

5

(a) $\dfrac{8}{27}$

(b) $P(SSR) + P(SRS) + P(RSS) = \dfrac{1}{3}\cdot\dfrac{3}{4}\cdot\dfrac{1}{4} + \dfrac{1}{3}\cdot\dfrac{1}{4}\cdot\dfrac{1}{3} + \dfrac{2}{3}\cdot\dfrac{1}{3}\cdot\dfrac{3}{4} = \dfrac{37}{144}$

(c) $P(SRR) + P(RSR) + P(RRS) = \dfrac{1}{3}\cdot\dfrac{1}{4}\cdot\dfrac{2}{3} + \dfrac{2}{3}\cdot\dfrac{1}{3}\cdot\dfrac{1}{4} + \dfrac{2}{3}\cdot\dfrac{2}{3}\cdot\dfrac{1}{3} = \dfrac{7}{27}$

(d) $P(RRR) + P(RSR) + P(SRR) + P(SSR) = \dfrac{2}{3}\cdot\dfrac{2}{3}\cdot\dfrac{2}{3} + \dfrac{2}{3}\cdot\dfrac{1}{3}\cdot\dfrac{1}{4} + \dfrac{1}{3}\cdot\dfrac{1}{4}\cdot\dfrac{2}{3} + \dfrac{1}{3}\cdot\dfrac{3}{4}\cdot\dfrac{1}{4} = \dfrac{203}{432}$

(e) $\dfrac{P(RRR) + P(RSR)}{P(RRR) + P(RSR) + P(SRR) + P(SSR)} = \dfrac{19/54}{203/432} = \dfrac{152}{203}$

6 $\dfrac{1}{3}$

7 $\dfrac{1/270725}{6961/270725} = \dfrac{1}{6961}$

8

(a) $\dfrac{2}{7}$

(b) $\dfrac{2}{3}$

(c) $\dfrac{1}{5}$

9 $\dfrac{11}{32}$

Topic 13

Probability Distribution

Functions

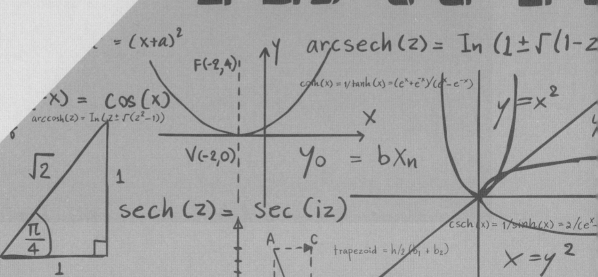

13.1 Discrete Random Variable

A random variable represents the possible outcomes of any experiment. A discrete random variable x results in a discrete probability distribution, where the following formula is used.

$X = x_i$	x_1	x_2	$\cdots$	x_n
$P(X = x_i)$	$P(X = x_1)$	$P(X = x_2)$	$\cdots$	$P(X = x_n)$

where $P(X = x_i)$ is a probability mass function.

Probability Distribution for Discrete Random Variable

Mean or Expectation $= \sum x \times P(x)$

Variance $= \sum x^2 \times P(x) - \mu^2$

where variance is deduced by $Var(X) = E(X - \mu)^2 = \sum(x_i - \mu)^2 p_i = \sum(x_i^2 p_i) - \mu^2$. Furthermore, a game is said to be fair if $E(X) = 0$.

1 A sixteen-sided spinner consists of a regular 16-gon with a pin through its center perpendicular to its plane. It has two regions labeled with 1, three regions with 2, four regions with 3, four regions with 4 and three regions with 5. Find the mean and variance of the score which results from one spin of the spinner.

2 Find the value of k in the following probability distribution. Deduce the mean and the variance of x.

x	1	2	3	4	5
$p(X = x)$	k	$2k$	$3k$	$2k$	k

3 Investigate how the following probability distribution could go wrong or is already false.

(a)

x	1	2	3	4	5
$p(X = x)$	k	$2k$	$3k$	$2k$	k

(b) $P(X = x) = \begin{cases} \dfrac{4-x}{5} & \text{for } x = 1,2,3,4,5 \\ 0 & \text{otherwise} \end{cases}$

4 A discrete random variable X satisfies the following probability distribution.

x	1	2	3
$p(X = x)$	k	$0.8 - k$	0.2

Given that the mean of the distribution is equal to 1.7, determine the value of k.

13.2 Binomial Distribution

A situation dealing with two possible outcomes, labeled 'success/failure' is called a binomial situation. In order to satisfy a binomial distribution, the probability of a success(or a failure) should constant. If p is the probability of a success, then the probability of getting r number of successes in n trials is given by

$$P(sss...ssrrr..rrr) = \frac{n!}{r!(n-r)!}p^r(1-p)^{n-r}$$

If $X \sim B(n,p)$[1], then it can be shown that

$$E(X) = np$$
$$Var(X) = npq$$

$\boxed{5}$ Suppose 60% is the unchanged chance that a day will be rainy. Find the chance of

(a) 4 rainy days in a week

(b) less than 4 rainy days in a week

(c) Hence, what is the expected number of rainy days in a week and what is its variance?

---

[1]The notation implies that a random variable X follows the binomial distribution model with n trials with the success probability of p.

6 On average a train arrives late on 15 journeys out of 100. Next week, Bob will make 5 train journeys in total. Let X denote the number of times his train will be late.

(a) State one assumption which must be made for X to be modeled by a binomial distribution.

(b) Find the probability that a train will be late on all of the 5 journeys.

(c) Find the probability that a train will be late on 2 or more out of the 5 journeys.

7 Given that $X \sim B(7, 0.6)$, find $P(X = 5)$. (The notation stands for a random variable X following the binomial distribution model with 7 trials with the success probability of 0.6.)

8 Suppose that 30% of dogs have skin problems. Bob is currently raising 5 dogs. Let X denote the number of his dogs with skin problems.

(a) State one assumption that must be satisfied for X to be modeled on a binomial distribution.

(b) Find the chance that (i) exactly 3 of his dogs have skin problems, and (ii) no more than 3 dogs have skin problems.

(c) How many of his dogs are expected to have skin problems?

(d) Write down $E(X)$ and $Var(X)$.

9 Given that $X \sim B(n, \frac{1}{4})$, and $E(X) = 200$, find $Var(X)$.

10 Given that $X \sim B(n, p)$, if $E(X) = 20$, $Var(X) = 10$, find the values of n and p.

13.3 Normal Distribution

Normal distribution is an extension of binomial distribution, i.e., a change from discrete random variables to continuous random variables. Most of the natural behavior follows normal distribution, i.e., showing a strong clustering around the mean(one central value). Nonetheless, this is an optimal form of distribution. None of what we observe actually fit into this model. This is the reason why we use approximation in statistics. A normal distribution has the following shapes, which can also be written as $X \sim N(\mu, \sigma^2)$.

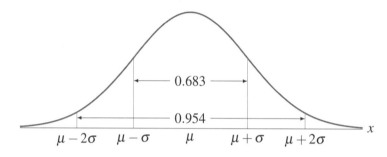

The standard deviation shows an interesting property such that approximately 68% of the values lie within one standard deviation away from the mean and about 95(or 96)% of the values lie within two standard deviations from the mean.

In order to use normal tables, we use $z = \dfrac{x - \mu}{\sigma}$ where z is the number of standard deviations away from the mean, also known for Z-score.

11 (Part 1. Graph Question)

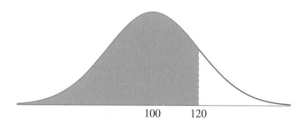

Find the area of the colored region under the curve, using that $\sigma = 10$ and normalcdf(-1e99,2,0,1)≈0.977.

11 (Part 2. Non-graph Question) Given $X \sim N(50, 16)$, find the probability of being less than 47, where $N(50, 16)$ means that the mean is 50 and the variance is 16.

12

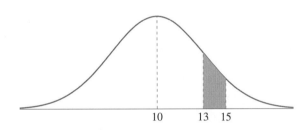

Find the probability of getting the colored portion on the normal distribution using calculator where $\sigma = 4$ where σ is the standard deviation. (Hint: Use your normalcdf function from your calculator.)

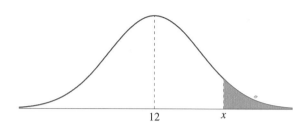

If the colored portion is 11%, then find x using the inverse normal probabilities, where $\sigma = 4$. Use that invNorm(0.89,0,1)≈1.227.

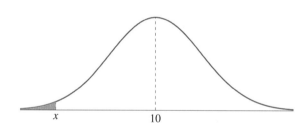

If the colored portion is 5%, then find x using the inverse normal probabilities, where $\sigma = 2$. Make sure you check that the answer is less than 10, which is the mean value. Use invNorm(0.05,0,1)≈-1.645.

1 Suppose 5 out of 100 factory items constantly turn out to be problematic. If there are 20 items freshly made in the factory, find the chance of

(a) containing exactly 6 problematic items.

(b) containing less than 2 problematic items.

2 Find the value of k for the following probability distribution, and find the mean value (or expected value).

x	1	2	3	4	5
$p(X = x)$	k	$2k$	$3k$	$4k$	$5k$

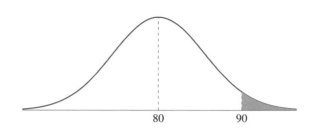

Math scores at your school is normally distributed with the mean score of 80. If 3% of the scores have a value greater than 90, find the standard deviation. Use that invNorm(0.97,0,1)≈ 1.881.

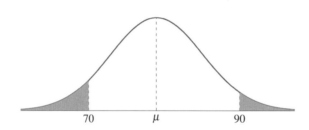

Find μ and σ by using simultaneous equations, if the colored portion on the left corresponds to 2.5% and the colored portion on the right to 0.5%. Use invNorm(0.025,0,1)≈-1.96 and invNorm(0.995,0,1)≈2.58.

5 If $X \sim N(50, 16)$, what is the mean and standard deviation of this normal distribution?

6 Given that $X \sim N(10, 2.25)$, find $P(X > 12)$. Use that normalcdf(-1E99,8,10,1.5)≈0.091.

7 If $X \sim N(60, 25)$ and $P(X > a) = 0.24$, find the value of a. Use invNorm$(0.76, 0, 1) \approx 0.71$.

8 If $X \sim N(45, \sigma^2)$ and $P(X > 51) = 0.28$, find σ. Use invNorm$(0.72, 0, 1) \approx 0.58$.

1

(a) $\dfrac{20!}{6!14!} \times \left(\dfrac{1}{20}\right)^{6} \times \left(\dfrac{19}{20}\right)^{14}$

(b) $\dfrac{20!}{0!20!} \times \left(\dfrac{1}{20}\right)^{0} \times \left(\dfrac{19}{20}\right)^{20} + \dfrac{20!}{1!19!} \times \left(\dfrac{1}{20}\right)^{1} \times \left(\dfrac{19}{20}\right)^{19}$

2 $k = \dfrac{1}{15}$, and $E(X) = \dfrac{11}{3}$.

3 $\sigma = \dfrac{10}{1.881} \approx 5.32$

4 $\mu = 78.6344$, $\sigma = 4.40529$

5 The mean is 50 and the standard deviation is 4.

6 $P(X > 12) = 0.091$.

7 $a = 63.55$

8 $\sigma = \dfrac{6}{0.58} \approx 10.3448$

Algebra 2에서 공부한 내용들을 더 잘 기억하기 위해서 저자가 알려주는 Tip!

1. 심리학적으로 초과학습은 많이 연구된 현상으로, 필요한 것 이상으로 연습을 더 하면 기억에 오래 남는다.

2. 문제 해설을 보기 전에 틀리더라도 직접 풀어보는 것이 기억에 제일 남는다.

3. 겉보기에 새롭고 어려워 보여도, 그 핵심이 무엇일까 고민하자!

4. Algebra 2에서 배우는 개념들이 어떻게 작용될까, 어떤 방식으로 쓰일까 고민하며 학습하자.

5. Algebra 2에서 다루는 내용은 '실수'에 관한 내용이므로, '실수'의 성질에 대해 고민하며 학습하자.

6. 함수를 공부할 때, Domain / Range 만 학습하지 말고, 위 함수가 1-대-1 함수일지 생각해보고, 이러한 것들이 1-대-1 대응으로 문제풀이에 어떻게 적용될지 고민하면서 학습하자.

7. 개념편에 나온 내용들은 고등학교 과정에서 풀어낼 문제들의 Foundation이 되는 내용이므로, 온전히 나의 것으로 만들자.

Solution Manual

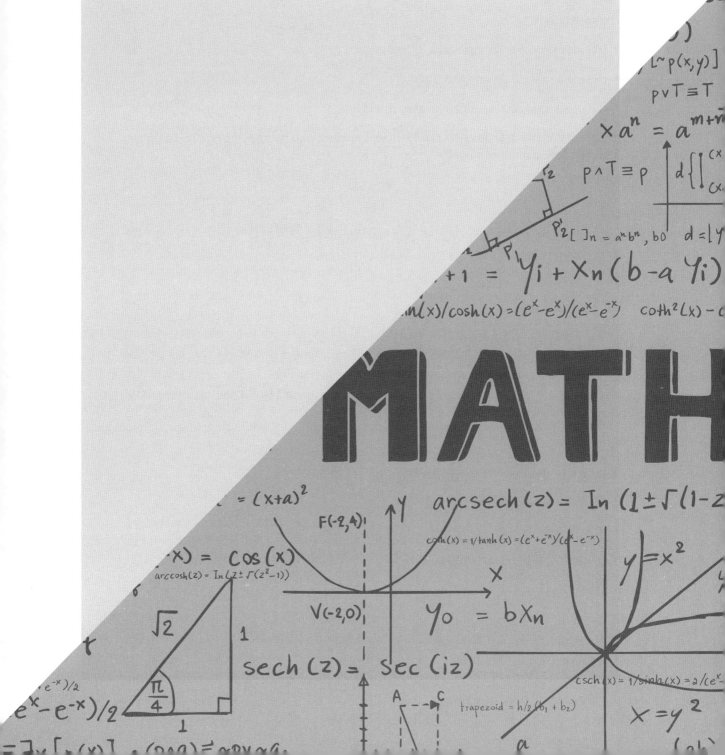

1 The answer is (D).

- (A) $x+0 = 0+x = x$ for any real x.

- (B) $x+y = y+x$ for any real x and y.

- (C) $x+(-x) = (-x)+x = 0$ for any real x.

- (D) $a+(b+c) = (a+b)+c$ for any real a, b, and c.

2 The answer is (C).

- (A) $a(bc) = (ab)c$ for any real a, b, and c.

- (B) $a+b = b+a$ for any real a and b.

- (C) $a(b+c) = ab+ac$ for any real a, b, and c.

- (D) Adding two real numbers results in real number, and multiplying two real numbers results in real number as well.

3

(a) $-\sqrt{2} \in \mathbb{R} \subset \mathbb{C}$.

(b) $-3 \in \mathbb{Z} \subset \mathbb{Q} \subset \mathbb{R} \subset \mathbb{C}$.

(c) $\dfrac{3}{4} \in \mathbb{Q} \subset \mathbb{R} \subset \mathbb{C}$.

(d) $\sqrt{-3} \in \mathbb{C}$.

(e) $\sqrt[3]{-8} \in \mathbb{Z} \subset \mathbb{Q} \subset \mathbb{R} \subset \mathbb{C}$.

4 Since $\sqrt{1} < \sqrt{2} < \sqrt{3} < \sqrt{4} < \sqrt{5}$, we get $1 < \sqrt{2} < \sqrt{3} < 2 < \sqrt{5}$.

5 (a) Since $1 < \sqrt{3} < 2$, and the only dot that is between 1 and 2 is (C), the answer must be (C).

(b) First categorize the answer choices. A is in the leftside of -0.5, while other answer choices are in the rightside of -0.5.

- $|-2-(-0.5)| = -0.5-(-2) = 1.5$

- $|B-(-0.5)| \approx 0.3-(-0.5) = 0.8$

- $|C-(-0.5)| \approx 1.25-(-0.5) = 1.75$

- $|D-(-0.5)| \approx 2.5-(-0.5) = 3$

Hence, the answer is (B).

$\boxed{6}$ Let x be the midpoint of -3 and 7. Then, $-3 < x < 7$. Hence,

$$|x-(-3)| = |7-x|$$
$$x+3 = 7-x$$
$$2x = 4$$
$$x = 2$$

Hence, the answer is (C).

$\boxed{7}$ The answer is (D).

- (A) $|2-(-2)| = 2-(-2) = 4$.
- (B) $|\frac{4}{3}-(-2)| = \frac{4}{3}+2 = \frac{10}{3}$.
- (C) $|-\frac{1}{2}-(-2)| = -\frac{1}{2}-(-2) = -\frac{1}{2}+2 = \frac{3}{2}$.
- (D) $|3-(-2)| = 3+2 = 5$.

$\boxed{8}$ Let's casework. Assume that $x < -1$. Then, $|x-(-1)| = (-1)-x = 4$, so $x = -5$. On the other hand, assume $x > -1$. Then, $|x-(-1)| = x+1 = 4$, so $x = 3$. Therefore, $x = 3$ or -5.

$\boxed{9}$ Let x be the smallest number. Then, $x+(x+1)+(x+2) = 72$, so $x = 23$. Hence, the three numbers are 23, 24, and 25. If we let x be the middle number, then $(x-1)+x+(x+1) = 3x = 72$, so $x = 24$. Nonetheless, the three numbers are still 23, 24, and 25.

$\boxed{10}$ Let $x = 3y-4$ and $x+y = 44$. Since we have two variables with two equations, we can always find the solution. Hence, $4y-4 = 44$, so $y = 12$ and $x = 32$.

$\boxed{11}$

$$\left(x \text{ employees} \times \frac{\$2500}{1 \text{ employee}} + \$3000 \right) / 1 \text{ month} \times \frac{12 \text{ months}}{1 \text{ year}}$$

$\boxed{12}$ $(x+y)^2 - xy = x^2 + 2xy + y^2 - xy = x^2 + xy + y^2$.

$\boxed{13}$ Let x be the age of John and y be that of Jane. Then, $x = y+7$ and $x+5 = 2(y+5)$. Hence, $y = 2$ and $x = 9$.

14 This is a typical unit conversion question.

$$60t = 1200 + 20(t - 10)$$
$$40t = 1000$$
$$t = 25 \text{(minutes)}$$

15

$$\sqrt{x} + 5 = 9$$
$$\sqrt{x} = 4$$
$$x = 16$$
$$x + 2 = 18$$

16 Given $4x + 10y = 60$, let's casework on y-values.

1. if $y = 1$, then $4x + 10 = 60$, so x is not an integer.

2. if $y = 2$, then $4x + 20 = 60$, so $x = 10$.

3. if $y = 3$, then $4x + 30 = 60$, so x is not an integer.

4. if $y = 4$, then $4x + 40 = 60$, so $x = 5$.

5. if $y = 5$, then $4x + 50 = 60$, so x is not an integer.

17 Since $3x - 3 = 21$, so $3x = 24$. Hence, $x = 8$. Therefore, $8 + \frac{1}{2}(x) = 8 + \frac{1}{2}(8) = 8 + 4 = 12$.

18

$$x + 2 < 2x - 1$$
$$2 + 1 < 2x - x$$
$$3 < x$$

19 First, $c \geq 0$ and $s \geq 0$, since there cannot be any negative number of cookies(or fruit salads). Second, $5c + 8s \leq 40$.

20

$$-5 < 3x - 2 < 11$$
$$-5 + 2 < 3x < 11 + 2$$
$$-3 < 3x < 13$$
$$-1 < x < \frac{13}{3}$$

21 First, $2x + 3 < -7$ implies $x < -5$. Second, $-2x - 1 < 7$ implies $-4 < x$. Since the inequalities are disjunctive, the solutions are $(-\infty, -5) \cup (-4, \infty)$ (in interval notation).

22 Assume $x < y$. Then, $|x - y| = y - x$. Hence, $|x - y| - (x - y) = y - x - x + y = 2y - 2x$. Therefore, the answer must be (C).

| 23 |

$$|x+1| = 3$$
$$x+1 = \pm 3$$
$$x = -1 \pm 3$$
$$x = -4 \text{ or } 2$$

and -1 is the midpoint.

| 24 |

(a) There is no such x satisfying $|2x-5| < 0$.

(b) $|2x-5| = 0$ at $x = \dfrac{5}{2}$.

(c) $|2x-5| > 0$ for all real number $x \neq \dfrac{5}{2}$.

| 25 |

$$|x-2| = 10$$
$$x-2 = \pm 10$$
$$x = 2 \pm 10$$
$$x = -8 \text{ or } 12$$

Hence, the sum of two possible values of x must be $4(= -8 + 12)$.

| 26 |

(a) Assume $x \geq 1$. Then, $|x-1| = x-1 = 2x+3$ must be solved by $x = -4$. This contradicts the original assumption, so there is no real x satisfying $|x-1| = 2x+3$ when $x \geq 1$.

(b) Assume $x < 1$. Then, $|x-1| = 1-x = 2x+3$ must be solved by $x = -\dfrac{2}{3}$. This matches with the assumption, so $x = -\dfrac{2}{3}$ must be the solution.

| 27 |

Since $|3x-4| = -2x+1$, either $3x-4 = -2x+1$ or $3x-4 = 2x-1$. Hence, $x = 1$ or $x = 3$. Substitute these two values into the original equation. At $x = 1$, $|3(1)-4| = -2(1)+1$, which is not true. At $x = 3$, $|3(3)-4| = -2(3)+1$, which is not true. There is no real solution to this equation.

| 28 |

$$6\frac{1}{4} < h < 6\frac{3}{4} \implies \left|h - 6\frac{1}{2}\right| < \frac{1}{4}$$

| 29 |

$$400 < l < 410 \implies |l - 405| < 5$$

1 The correct answer is (B) because $f(3) = -1$ and if $f(3) = -2$, then 3 is connected to -1 and -2, which means that this will not pass the vertical line test.

2 The domain of the relation given by the expression must be all x-values counted once. Hence, the answer is (C).

3 $f(2x+1) = 3(2x+1)^2 + 4 = 3(4x^2 + 4x + 1) + 4 = 12x^2 + 12x + 7$.

4 The value of m must be 2.

5 $f(1) + f(2) + f(3) + f(4) = 1^1 + 2^2 + 3^3 + 4^4 = 1 + 4 + 27 + 256 = 288$.

6 Let $D = k \times S$ where k is the constant of variation. Then, $k = 40$ by plugging $(D, S) = (240, 6)$. Since 100 ft equals 1200 inches, we get $1200 = 40S$. So, $S = 30$ strokes.

7 Let $S = kW$ where k is the constant of variation. Plugging the given values results in $k = 7.2$. Hence, plugging $w = 6.4$ gives $S = \dfrac{1152}{25}$.

8

(a) $T = k \times g$. Substituting $(T, g) = (540, 12)$, we get $k = \dfrac{540}{12} = \dfrac{135}{3}$. Hence, $T = \dfrac{135}{3}g$.

(b) Let x be the number of miles that owner travels on 9 gallons of gas. Hence,

$$9 \text{ gallons} \times \frac{135 \text{ miles}}{3 \text{ gallons}} = 405 \text{ miles}$$

9 Let $C = k \times m$ where C is the cost and m is the number of minutes. Then, $k = \dfrac{3}{50}$. Hence, $1.32 \times \dfrac{50}{3} = 22$ minutes.

10 -4 in the equation can be represented by

$$\frac{-4 \text{ hours}}{+1°C}$$

Hence, the answer is (A).

11 At $t = 0$, the price of the car must be finalized. That means that the final auction price of the car. Hence, the answer is (B).

12 Given two points $(-3, 5)$ and $(6, 8)$, the line equation must be $y = \dfrac{1}{3}(x - 6) + 8$. Hence, the answer must be (A), since $y = 7$ at $x = 3$.

13 By drawing the line passing through $(3, -7)$ and $(3, 2)$, we get a vertical line $x = 3$. Therefore, the answer is (B).

14 Substituting $(2,8)$ results in $b = -4$. Hence, $y = mx + b = mx - 4$. Thus, $8 = 2m - 4$ implies $2m = 12$, so $m = 6$. The answer is (C).

15 Since the function $f(x)$ is either increasing, decreasing or constant, the answer must be (D).

16 The line parallel to the line with the slope of -2 has the slope of -2. So, the answer must be (C).

17 The line passing through $(-1,1)$ with the slope of $\dfrac{2}{3}$ is $y = \dfrac{2}{3}(x - (-1)) + 1 = \dfrac{2}{3}(x+1) + 1$, which is equivalent to (A).

18 The line perpendicular to l has the slope of -2. If it passes through $(1,5)$, then the equation must be $y = -2(x - 1) + 5 = -2x + 7$.

19

$$0 < \frac{a}{b} < 1$$
$$0 < -\frac{e}{d} < 1$$
$$0 > \frac{e}{d} > -1$$
$$\frac{d}{e} < -1$$

20 Given $y = f(2x - 2) + 2 = f(2(x-1)) + 2$, $(6,4)$ is shrunk to $(3,4)$, then shifted 1 unit right and 2 units up. Hence, the answer is (B).

21 Given $y = f(2x)/2 + 2$, $(6,4)$ is shrunk to $(3,2)$ and shifted 2 units up, so the answer must be (D).

22 Given $y = f(2 - x) + 2 = f(-(x - 2)) + 2$, $(6,4)$ is reflected about the y-axis, then shifted 2 units right and 2 units up, so the answer must be (D).

23 Since $f(x)$ is reflected about the x-axis, the equation turns into $y = -f(x)$. Afterwards, the new equation turns into $y = -f(x - 3)$ because the graph is shifted 3 units right. Hence, the answer must be (C).

24

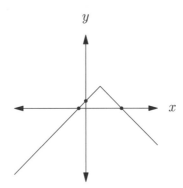

25

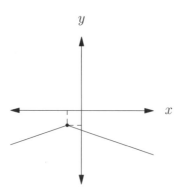

26 According to the four listed transformations, $y = |x|$ is transformed into $y = \frac{1}{3}|x+2| + 4$. Hence, the answer is (C).

27

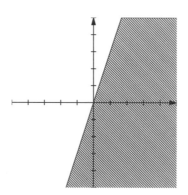

28

$$\begin{cases} v \geq 0 \\ c \geq 0 \\ 8v + 4c \geq 120 \end{cases}$$

where v is the number of vans and c is the number of cars.

29 Let x be the number of brand #1 cleaners and y be the number of brand #2 cleaners. Then, it is obvious that $x \geq 0$ and $y \geq 0$. Since the salesperson's goal is to have at least $\$1,800$ worth of sales, $150x + 200y \geq 1800$ at the same time.

$\boxed{1}$ Let t be the time and y be the number of flyers. Then, $y = 80 + 6t$ and $y = 100 + 4t$. At $t = 0$, then you and your partner send 180 flyers altogether.

$\boxed{2}$

(a) $\begin{cases} c(d) &= 25d \\ c(d) &= 20d + 30 \end{cases}$

(b) $C_x = 25(7) = 175$ and $C_y = 20(7) + 30 = 170$. In this case, we should choose Y because it is cheaper.

$\boxed{3}$ The answer is (B) because $\dfrac{6}{4} = \dfrac{3}{2} \neq \dfrac{12}{4}$.

$\boxed{4}$ The answer is (B). Let x be the amount of time and y be the number of scarves. Then, $y = 10 + 8x$ and $y = 19 + 5x$ at the same time. Hence, $10 + 8x = 19 + 5x$, so $x = 3$ (minutes).

$\boxed{5}$

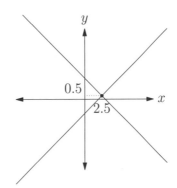

$\boxed{6}$ Let x be the price of orange and y that of grapefruit. Then, we get $8x + y = 4.6$ and $6x + 3y = 4.8$. Substituting $y = 4.6 - 8x$ into the second equation, we get $x = 0.5$ and $y = 0.6$ in dollars.

$\boxed{7}$

(a) $\begin{cases} f + s &= 15 \\ 2f - 3s &= 5 \end{cases}$ (b) $f = 10$ and $s = 5$ (apartments)

$\boxed{8}$ Let x be the number of books and y be the amount of dollars. Then, $y = 15.49x$ and $y = 13.99x + 6$. Eliminating y, we get $x = 4$ (books).

$\boxed{9}$ Let x be the number of bats and y be the number of gloves. Then,

$$\begin{cases} 20x + 12y &= 1040 \\ 25x + 16y &= 1350 \end{cases}$$

Hence, $x = 22$ and $y = 50$.

$\boxed{10}$ Since $3y = 2x$ and $2(3y + 2x) = 72$, we get $x = 9$ and $y = 6$.

$\boxed{11}$ Let x be the number of spiral notebooks and Y that of 3-ring notebooks. First, $x \geq 0$ and $y \geq 0$. Also, $x + y \geq 6$ and $2x + 5y \leq 20$. Hence, the system of inequalities we look for is

$$\begin{cases} x \geq 0 \\ y \geq 0 \\ x + y \geq 6 \\ 2x + 5y \leq 20 \end{cases}$$

$\boxed{12}$

(a) 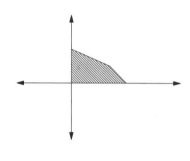   (b)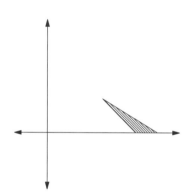

$\boxed{13}$

(a)   (b)

$$\begin{cases} x \geq 0 \\ y \geq 0 \\ 6x + 30y \leq 600 \\ x + y \leq 60 \end{cases}$$

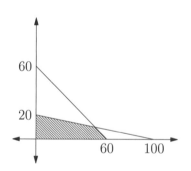

(c) We need to optimize $2.5x + 7.5y$ where critical points are $(0, 20)$, $(0, 0)$, $(60, 0)$ and $(50, 10)$. At $(50, 10)$, the maximum income becomes 200 dollars.

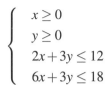

(a)

(b)

$$\begin{cases} x \geq 0 \\ y \geq 0 \\ 2x + 3y \leq 12 \\ 6x + 3y \leq 18 \end{cases}$$

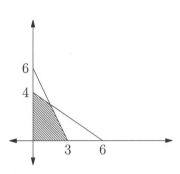

(c) We need to optimize $18x + 12y$ where critical points are $(0,0)$, $(3,0)$, $(0,4)$ and $(1.5,3)$. At $(1.5,3)$, the maximum profit becomes 63 dollars.

15

(a)

(b)

$$\begin{cases} x \geq 0 \\ y \geq 0 \\ x + y \leq 300 \\ x + 3y \leq 360 \end{cases}$$

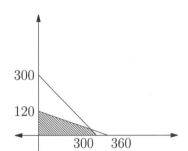

(c) We need to optimize $300x + 600y$ where critical points are $(0,0)$, $(0,120)$, $(300,0)$ and $(270,30)$. At $(270,30)$, the maximum revenue becomes 99000 dollars.

16 The correct answer is (D) because $2(0) + 3(3) + 3 = 12$.

17 The distance between $(0,0,0)$ and $(1,3,4)$ is $\sqrt{1^2 + 3^2 + 4^2} = \sqrt{26}$.

18 The answer is (C). If two of three planes are perpendicular, then there is one point of intersection. Otherwise, it must form a line.

19 Use the last equation to substitute $y = z + 5$. Hence, $z = -4$ and $x = -z = -(-4) = 4$. Also, $y = (-4) + 5 = 1$.

20 Add the first two equations to get $x = 0$. Hence, $y - z = -1$ and $-2y + z = 3$. Adding the two equations results in $-y = 2$. Therefore, $y = -2$. Hence, $z = -1$.

21 The answer is (A) because there are three rows and four columns.

22 Since $2 \begin{bmatrix} 2 & 4 \\ 2 & -1 \end{bmatrix} - 3 \begin{bmatrix} 1 & -1 \\ -3 & 2 \end{bmatrix} = \begin{bmatrix} 3 & 11 \\ 13 & -8 \end{bmatrix}$. Hence, the sum of entries is 19. The answer is (C).

23 $\begin{bmatrix} 1 & 0 \\ 0 & -1 \end{bmatrix} \times \begin{bmatrix} 2 & 3 \\ -1 & 2 \end{bmatrix} = \begin{bmatrix} 2 & 3 \\ 1 & -2 \end{bmatrix}.$

24

(a) The inverse of A is $\dfrac{1}{7} \begin{bmatrix} 3 & -1 \\ 1 & 2 \end{bmatrix}.$

(b) The value of n is 2.

25 $\begin{bmatrix} x \\ y \end{bmatrix} = \dfrac{1}{7} \begin{bmatrix} 3 & 1 \\ 2 & 3 \end{bmatrix} \begin{bmatrix} 4 \\ 2 \end{bmatrix} \implies \begin{bmatrix} x \\ y \end{bmatrix} = \dfrac{1}{7} \begin{bmatrix} 3(4) + 1(2) \\ 2(4) + 3(2) \end{bmatrix} \implies \begin{bmatrix} x \\ y \end{bmatrix} = \begin{bmatrix} 2 \\ 2 \end{bmatrix}.$

26 The determinant of $\begin{bmatrix} 2 & 1 \\ 4 & 2 \end{bmatrix}$ is $2(2) - 1(4) = 0.$

1

(a)   (b)   (c)

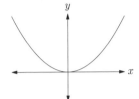   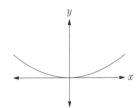

2

(a) $(-3, 2)$ (b) $x = -3$ (c) $y = 2$ (d) $\{y \in \mathbb{R} : y \leq 2\}$

3

(a) The graph is not dilated, but it is reflected about the x-axis, translated 1 unit left and 3 units down.

(b) The graph is not reflected about any line, but it is vertically shrunk by 2 units and translated 1 unit right and 1 unit down.

4

(a) $y = \dfrac{1}{2}(x-3)^2 + 1$ (b) $y = -\dfrac{8}{9}(x - \dfrac{1}{2})^2 + 1$

5 Since $A(x) = -4x^2 + 40x = -4x(x - 10)$, the vertex is located at the midpoint of 0 and 10. Hence, at $x = 5$, $A(5) = -4(5)(5 - 10) = 100$. The area must be 100 square inches.

6

(a) The vertex is located at $(-2, 3)$. The axis of symmetry, hence, is $x = -2$, and the minimum value of the function is 3. Therefore, its range equals $[3, \infty)$.

(b) The vertex is located at $(1, 2)$. The axis of symmetry is $x = 1$, and its maximum value is 2. Therefore, the range equals $(-\infty, 2]$.

(c) The vertex is located at $\left(\dfrac{1}{4}, \dfrac{7}{8}\right)$. Hence, the axis of symmetry is $x = \dfrac{1}{4}$, and its minimum value is $\dfrac{7}{8}$. Thus, the range equals $[\dfrac{7}{8}, \infty)$.

7

(a) $y = (x - 3)^2$ (b) $y = (x - 2)^2 - 4$

(c) $y = 2\left(x - \dfrac{5}{2}\right)^2 - \dfrac{21}{2}$ (d) $y = -3\left(x - \dfrac{1}{3}\right)^2 + \dfrac{4}{3}$

(a) $P(n) = -\dfrac{1}{50}(n-85)^2 + 128.5$. Hence, 85 books.

(b) 128 thousands and 5 hundred dollars

9 Since $ax^2 - 10x + b = a(x-5)^2 + 2$, then $ax^2 - 10x + b = ax^2 - 10ax + (25a+2)$. Hence, $a = 1$ and $b = 27$.

10

(a) $(x-3)(x-7)$

(b) $(x-4)(x-8)$

(c) $-(x-5)(x-7) = (5-x)(x-7)$

(d) $-(y-6)(y+9) = (y-6)(-y-9)$

(e) $(x+12)(x-5)$

(f) $(x-3)(x-5)$

11

(a) $(5x-2)(x-3)$

(b) $(2x+1)(x-5)$

(c) $(3x-1)^2$

(d) $(x-7)(x+7)$

(e) $(3x+4)(x+2)$

(f) $(x+6)^2$

(g) $(2x+3)^2$

(h) $2(x-5)(x+5)$

12

(a) $x^2 - 30x + (-15)^2 = (x-15)^2$

(b) $x^2 + 5x + \left(\dfrac{5}{2}\right)^2 = \left(x+\dfrac{5}{2}\right)^2$

(c) $x^2 - \dfrac{1}{3}x + \left(\dfrac{1}{6}\right)^2 = \left(x-\dfrac{1}{6}\right)^2$

(d) $2x^2 + 4x + 2 = 2(x+1)^2$

(e) $20x^2 + 10x + \dfrac{5}{4} = 20\left(x+\dfrac{1}{4}\right)^2$

(f) $4x^2 - 10x + \dfrac{25}{4} = 4\left(x-\dfrac{5}{4}\right)^2$

13

(a) $k = 24$

(b) $k = 12$

(c) $k = 10$

(d) $k = 24$

14 Since $d(t) = -t^2 + 6t + 23 = -(t-3)^2 + 32$, the maximum depth of water is 32 feet at $t = 3$ in hours.

15

(a) $x = \dfrac{4 \pm 2\sqrt{5}}{2} = 2 \pm \sqrt{5}$

(b) $x = \dfrac{5 \pm \sqrt{13}}{2}$

16

(a) There is no real zero (but two complex zeros) because the discriminant is negative.

(b) There are two real zeros because the discriminant is positive.

(c) There is no real zero (but two complex zeros) because the discriminant is negative.

17

The graph of $y = x^2 + kx + 4$ is tangent to the x-axis if and only if the discriminant is 0. Hence, $k^2 - 4(4) = 0$. Therefore, $k = \pm 4$.

18

(a) $(-3 + 5i) + (2 - i) = (-3 + 2) + (5i - i) = -1 + 4i$

(b) $(1 + i)(4 - 2i) = 1(4) - 2i + 4i + 2 = 6 + 2i$

(c) $(3 - i)^2 = 9 - 6i + i^2 = 8 - 6i$

(d) $(-1 + 3i)^2 = (-1)^2 - 2(3i) + (3i)^2 = 1 - 6i - 9 = -8 - 6i$

(e) $1 + \sqrt{-3} - \sqrt{-27} = 1 + \sqrt{3}i - 3\sqrt{3}i = 1 - 2\sqrt{3}i$

(f) $(2i - 1)(3i) = 6i^2 - 3i = -6 - 3i$

19

(a) $\dfrac{5 + 2i}{3i} = \dfrac{5i - 2}{3i^2} = \dfrac{5i - 2}{-3} = \dfrac{2 - 5i}{3}$

(b) $\dfrac{2i}{2 - i} = \dfrac{2i(2 + i)}{5} = \dfrac{4i - 2}{5}$

(c) $\dfrac{1 - i}{2 + i} = \dfrac{(1 - i)(2 - i)}{(2 + i)(2 - i)} = \dfrac{1 - 3i}{5}$

20

(a) $x^2 + 4 = (x - 2i)(x + 2i)$ 　　　　　　 (b) $4x^2 + 9 = (2x + 3i)(2x - 3i)$

21

(a) $x^2 + 4x + 5 = 0$ can be turned into $(x + 2)^2 = -1$. Hence, $x = -2 \pm \sqrt{-1} = -2 \pm i$.

(b) $-x^2 + 4x - 10 = 0$ can be turned into $(x - 2)^2 = -6$. Hence, $x = 2 \pm \sqrt{6}i$.

(c) $2x^2 - 5x + 3 = 0$ can be factored into $(2x - 3)(x - 1)$ Hence, $x = \dfrac{3}{2}$ or 1.

(d) $-4x^2 - 6x + 1 = 0$ has the solution $x = \dfrac{-6 \pm \sqrt{52}}{8} = \dfrac{-6 \pm 2\sqrt{13}}{8} = \dfrac{-3 \pm \sqrt{13}}{4}$.

(e) $2x^2 - 2x + 6 = 0$ has the solution $x = \dfrac{1 \pm \sqrt{11}i}{2}$.

(f) $3x^2 - 2x + 5 = 0$ has the solution $x = \dfrac{2 \pm \sqrt{56}}{6} = \dfrac{2 \pm 2\sqrt{14}i}{6} = \dfrac{1 \pm \sqrt{14}i}{3}$.

(g) $4x^2 + 49 = 0$ can be factorized into $(2x + 7i)(2x - 7i) = 0$. Hence, $x = \dfrac{7i}{2}$ or $-\dfrac{7i}{2}$.

(h) $x^2 + x + 1 = 0$ has the solution $x = \dfrac{-1 \pm \sqrt{3}i}{2}$.

22

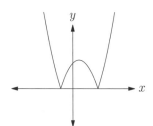

(a) The x-intercepts are -1 and 2.

(b) The relative maximum occurs at $x = \dfrac{1}{2}$.

23

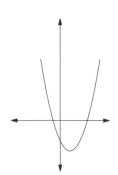

(a) The y-value of the endpoints is 6 for both.

(b) The y-value of the vertex is -3.

24

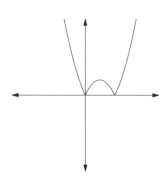

(a) The range of $f(x)$ is $[0, 3]$.

(b) $0 < k \le 1$.

$\boxed{25}$

(a) $A(x) = x(50 - 2x)$ where x is the width and $50 - 2x$ is the length.

(b) The vertex is located at $\left(\dfrac{25}{2}, \dfrac{625}{2} \right)$. Hence, the maximum area is $\dfrac{625}{2}$.

$\boxed{26}$ $3x^2 - 2x + 3 > 0$ because $(-2)^2 - 4(3)(3) < 0$.

$\boxed{27}$ $3x^2 - 4x \leq -1$ can be found by $3x^2 - 4x + 1 \leq 0$ for $\dfrac{1}{3} \leq x \leq 1$.

$\boxed{28}$ By discriminant method, $(2-k)^2 - 4(k^2 + 3k - 3) \geq 0$. Hence, $-3k^2 - 16k + 16 \geq 0$. Therefore, $3k^2 + 16k - 16 \leq 0$. Unfortunately, this cannot be factorized so we must use the quadratic formula to find out the smaller solution and the larger solution to $3k^2 + 16k - 16 = 0$.

Applying a quadratic formula, we get $k = \dfrac{-8 \pm 4\sqrt{7}}{3}$. Hence, the conjunctive inequality can be written as

$$\dfrac{-8 - 4\sqrt{7}}{3} \leq x \leq \dfrac{-8 + 4\sqrt{7}}{3}$$

$\boxed{29}$ $|x^2 - 3x - 7| \geq 3$ means that

- $x^2 - 3x - 7 \geq 3$

- $x^2 - 3x - 7 \leq -3$

First, $x^2 - 3x - 7 \geq 10 \implies x^2 - 3x - 10 \geq 0$. Second, $x^2 - 3x - 7 \leq -3 \implies x^2 - 3x - 4 \leq 0$. Solving these inequality, we get $x \leq -2$ or $-1 \leq x \leq 4$ or $5 \leq x$.

1 After expanding the whole expression, we get the degree of 2 and 2 terms. Therefore, the answer for (a) is 2 and that for (b) is 2, as well.

2

(a) $x^5 - x^4 + x^3$ has the degree of 5, and it is a trinomial.

(b) $-2x^3 + 9$ has the degree of 3, and it is a binomial.

3 The answer is (B). Since $x^2 - x + 3 > 0$ for all real x, $|x^2 - x + 3| = x^2 - x + 3$.

4

(a) The leading coefficient is positive because of 3.

(b) The degree is even because of 4.

(c) Hence, the end-behavior is up-up.

5

(a) The leading coefficient is negative because of -5.

(b) The degree is even because of 6.

(c) Hence, the end-behavior is down-down.

6

(a) The graph has no turning point, and the end-behavior is down-up.

(b) The graph has two turning points, and the end-behavior is up-down.

(c) The graph has two turning points, and the end-behavior is down-up

7

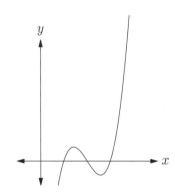

8 The degree of the polynomial function is 3 with the three layers of differences.

9 The degree of the polynomial function is 4 with the four layers of differences.

10

(a) $x^3 - 2x^2 + x = x(x-1)^2$ (b) $x^4 - 4x^2 + 4 = (x^2 - 2)^2$

11

(a) $x^3 + 7x^2 + 15x + 9 = (x+1)(x+3)^2$ (b) $-3x^3 + 18x^2 - 27x = -3x(x-3)^2$

12

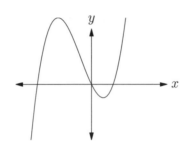

13

(a)   (b)

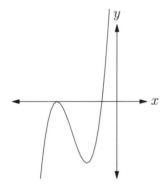   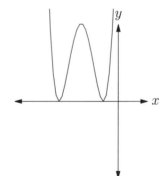

14

(a) $y = \alpha x^2(x+3)(x-5)$ where $\alpha \neq 0$

(b) $y = \alpha x(x-4)(x-2)(x+3)$ where $\alpha \neq 0$

(c) $y = \alpha(x-1)^2(x-2)^2$ where $\alpha \neq 0$

15 There are three zeros $x = 0, 3, -3$. In each case, the multiplicity is 1.

1

(a) $8x^3 - 27 = (2x-3)(4x^2+6x+9) = 0$. Hence, $x = \dfrac{3}{2}$ and $x = \dfrac{-3\pm3\sqrt{3}i}{4}$.

(b) $x^4 - 11x^2 + 10 = (x^2-10)(x^2-1) = (x-\sqrt{10})(x+\sqrt{10})(x-1)(x+1) = 0$. Hence, $x = \pm\sqrt{10}$ and ±1.

2

(a) $x^4 + 3x^2 - 4 = (x^2+4)(x^2-1) = 0$, so $x = \pm2i$ and $x = \pm1$.

(b) $2x^3 - 250 = 2(x^3-125) = 2(x-5)(x^2+5x+25) = 0$. Hence, $x = 5$ or $x = \dfrac{-5\pm5\sqrt{3}i}{2}$

3

(a)

$$3x^4 = 3x$$
$$3x^4 - 3x = 0$$
$$3x(x^3-1) = 0$$

Hence, $x = 0$ or $x = 1$ or $x = \dfrac{-1\pm\sqrt{3}i}{2}$.

(b)

$$x^4 - 13x^2 - 14 = 0$$
$$(x^2-14)(x^2+1) = 0$$

Hence, $x = \pm\sqrt{14}$ or $x = \pm i$.

(c)

$$x^4 - 16 = 0$$
$$(x^2-4)(x^2+4) = 0$$

Hence, $x = \pm2$ or $x = \pm2i$.

(d)

$$x^5 - x^3 - 12x = 0$$
$$x(x^4-x^2-12) = 0$$
$$x(x^2-4)(x^2+3) = 0$$

Hence, $x = 0, \pm2, \pm\sqrt{3}i$.

(e)

$$x^4 + 6 = 7x^2$$
$$x^4 - 7x^2 + 6 = 0$$
$$(x^2-6)(x^2-1) = 0$$

Hence, $x = \pm\sqrt{6}, \pm1$.

(f)

$$x^4 - x^2 - 132 = 0$$
$$(x^2-12)(x^2+11) = 0$$

Hence, $x = \pm2\sqrt{3}, \pm\sqrt{11}i$.

(g)

$$x^4 - 25 = 0$$
$$(x^2 - 5)(x^2 + 5) = 0$$

Hence, $x = \pm\sqrt{5}, \pm\sqrt{5}i$.

(h)

$$3x^4 - 24x^3 + 36x^2 = 0$$
$$3x^2(x^2 - 8x + 12) = 0$$
$$3x^2(x - 6)(x - 2) = 0$$

Hence, $x = 0, 2, 6$.

4

(a) $(x+3)(x-8) + 12 = x^2 - 5x - 12$

(b) $(x-1)(x^2 + 5x + 3) + 2 = x^3 + 4x^2 - 2x - 1$

(c) $(x+2)(x^2 - 7x + 17) - 32 = x^3 - 5x^2 + 3x + 2$

5
Applying the remainder theorem to the division, we get the remainder of 0. Hence, the quotient must be a factor of $x^3 + x^2 + x + 1$. By long division, we get $(x+1)(x^2 + 1) = x^3 + x^2 + x + 1$. Hence, $x^2 + 1$ is the quotient.

6

(a) $x^7 - 1 = (x-1)(x^6 + x^5 + x^4 + x^3 + x^2 + x + 1)$

(b) $x^4 + 3x^2 - 4x + 1 = (2x - 1)\left(\dfrac{x^3}{2} + \dfrac{x^2}{4} + \dfrac{13x}{8} - \dfrac{19}{16}\right) - \dfrac{3}{16}$

(c) $x^3 + 11x^2 - 13x + 24 = (x+1)(x^2 + 10x - 23) + 47$

7
Since $4x^3 + 4x^2 + x - 1 = (x+2)Q(x) + R$, then the remainder R equals -19.

8

(a) $x - 3$ is a factor of $x^3 + 3x^2 - 10x - 24$.

(b) $x + 4$ is a factor of $x^3 + 3x^2 - 10x - 24$.

(c) $x + 2$ is a factor of $x^3 + 3x^2 - 10x - 24$.

9
$x^3 + 3x^2 + 4x + 1$ has no positive factor because $x^3 + 3x^2 + 4x + 1 > 0$ if x is a positive number.

10

(a) $x^2 - 8x^2 + 17x - 10 = (x-5)(x^2 - 3x + 2)$.

(b) $x^3 - 5x^2 - 7x + 25 = (x-5)(x^2 - 7) - 10$.

(c) $x^3 + 2x^2 + 5x + 12 = (x+3)(x^2 - x + 8) - 12$.

11 By long division, we get $(2x-1)\left(\dfrac{1}{2}x^2 - \dfrac{7}{4}x + \dfrac{5}{8}\right) + \dfrac{13}{8}$. On the other hand, by synthetic division, we get $\left(x - \dfrac{1}{2}\right)\left(x^2 - \dfrac{7}{2}x + \dfrac{5}{4}\right) + \dfrac{13}{8}$.

12 The remainder is -1 either by synthetic division or remainder theorem.

13

(a) $1+i$ and $-\sqrt{5}$ are the other two roots.

(b) $2+\sqrt{3}$ and $-3-\sqrt{7}$ are the other two roots.

14

(a) Since all possible integer roots are factors of 6, we try $1,2,3,6$. Hence, $x^3 - 6x^2 + 11x - 6 = (x-1)(x-2)(x-3) = 0$. Therefore, $x = 1,\ 2$, and 3.

(b) Since all possible rational roots are $\pm 1, \pm\dfrac{1}{2}$, we try $1, \dfrac{1}{2}$. Try $x = 1$ in synthetic division. Then, we get $(x-1)(2x^2 - 3x + 1) = (x-1)^2(2x-1) = 0$.

15

(a) $x^3 - 4x^2 + 5x + 10 = (x+1)(x^2 - 5x + 10) = 0$, so $x = -1$ and $x = \dfrac{5 \pm \sqrt{15}i}{2}$

(b) $x^3 + x - 10 = (x-2)(x^2 + 2x + 5) = 0$, so $x = 2$ and $x = -1 \pm 2i$.

16

(a) $x^3 - 8x^2 + 9 = (x+1)(x^2 - 9x + 9) = 0$, so $x = -1$ and $\dfrac{9 \pm 3\sqrt{5}}{2}$.

(b) $x^3 + x^2 - 5x + 3 = (x-1)^2(x+3) = 0$, so $x = 1$ or -3.

(c) $x^3 + x^2 - 34x + 56 = (x-2)(x^2 + 3x - 28) = (x-2)(x-4)(x+7) = 0$, so $x = 2,\ 4$, and -7.

(d) $x^3 + x^2 - 27x + 25 = (x-1)(x^2 + 2x - 25) = 0$, so $x = 1$ or $-1 \pm \sqrt{26}$.

(e) $x^3 - 18x + 27 = (x-3)(x^2 + 3x - 9) = 0$, so $x = 3$ or $x = \dfrac{-3 \pm 3\sqrt{5}}{2}$.

(f) $x^3 - 5x^2 + x - 5 = (x^2 + 1)(x-5) = 0$, so $x = \pm i$ or 5.

17

(a) The number of positive real roots is 1. On the other hand, the number of negative real roots is either 2 or 0.

(b) The number of positive real roots is 1. On the other hand, the number of negative real roots is either 2 or 0.

18 Since $f(0) < 0$ and $f(1) > 0$, we should try a rational fraction between 0 and 1. This is because any polynomial function is continuous, meaning that the graph is always connected. The existence of real root between 0 and 1 is guaranteed by Intermediate Value Theorem.

1

(a) $\sqrt{169} = 13$ (b) $-\sqrt{36} = -6$ (c) $\sqrt[3]{0.008} = 0.2$ (d) $\sqrt[3]{-64} = -4$ (e) $\sqrt{0.16} = 0.4$

2

(a) $\sqrt[3]{216} = 6$ (b) $\sqrt[3]{-343} = -7$ (c) $\sqrt[3]{64} = 4$ (d) $\sqrt[3]{\dfrac{27}{1000}} = \dfrac{3}{10}$

3

(a) $\sqrt[4]{-81}$ is not real.   (b) $\sqrt[4]{256} = 4$ (c) $\sqrt[4]{64} = 2\sqrt{2}$ (d) $\sqrt[4]{81} = 3$

4

(a) $\sqrt[4]{\dfrac{x^4}{16}} = \dfrac{|x|}{2}$, since $\sqrt[4]{x^4} = |x|$ for any real x.

(b) $\sqrt[6]{(x-y)^6} = |x-y|$ for any real x and y.

(c) $\sqrt{8(a+b)^4} = \sqrt{8}\sqrt{(a+b)^4} = 2\sqrt{2}(a+b)^2$ for any real a and b.

(d) $\sqrt[4]{\dfrac{x^4}{64}} = \dfrac{|x|}{\sqrt[4]{64}} = \dfrac{|x|}{2\sqrt{2}} = \dfrac{|x|\sqrt{2}}{4}$

5

(a) Let r be the radius of a sphere. Then, $36\pi = \dfrac{4}{3}\pi r^3$. Hence, $r^3 = 27$. Since $r^3 - 27 = (r-3)(r^2+3r+9) = 0$, and $r^2+3r+9 = 0$ has no real solution, we may safely reach the conclusion that $r = \sqrt[3]{27} = 3$.

(b) Since $8V = \dfrac{4}{3}\pi(kr)^3$, then $k = 2$. Hence, the new radius is twice the original radius, which now has the measure of 6.

6

(a) $v^2 = 64 \times 10$, so $v = \pm 8\sqrt{10}$. Since the object is falling, the velocity must be negative. Hence, $v = -8\sqrt{10}$ feet per second. Velocity is a signed speed, whereas the speed has no positive or negative sign.

(b) To be realistic, the velocity in the question should be considered as the speed. Assuming that we consider the absolute value of the velocity, which is the definition of the speed, we get $v^2 = 64 \times 20$, so $v = 16\sqrt{5}$ if the object falls 20 feet. On the other hand, $v = 8\sqrt{10}$ if the object falls 10 feet. Hence, the difference between the two must be $16\sqrt{5} - 8\sqrt{10}$ feet per second.

7

(a)
$$\sqrt{25a^3} = \sqrt{25}\sqrt{a^3}$$
$$= 5a\sqrt{a}$$

(b)
$$\sqrt{20b^{10}} = \sqrt{20}\sqrt{b^{10}}$$
$$= 2\sqrt{5}b^5$$

(c)
$$\sqrt{5x^2}\sqrt{5y^3} = \sqrt{5^2}\sqrt{x^2}\sqrt{y^3}$$
$$= 5xy\sqrt{y}$$

(d)
$$7\sqrt{2} \times 3\sqrt{y^2} = 21\sqrt{2} \times \sqrt{y^2}$$
$$= 21\sqrt{2}y$$

(e)
$$\frac{\sqrt{35}}{\sqrt{7}} = \sqrt{\frac{35}{7}}$$
$$= \sqrt{5}$$

8

(a) $\sqrt{48x^3} = \sqrt{48}\sqrt{x^3} = 4\sqrt{3}x\sqrt{x} = 4x\sqrt{3x}.$

(b) $\sqrt[3]{216x^3y^4} = \sqrt[3]{216}\sqrt[3]{x^3}\sqrt[3]{y^4} = 6xy\sqrt[3]{y}.$

(c) $\sqrt{75a^3} = \sqrt{75}\sqrt{a^3} = 5\sqrt{3}a\sqrt{a} = 5a\sqrt{3a}.$

9

Part 1.

(a) $\sqrt{3} \cdot \sqrt{8} = \sqrt{24} = 2\sqrt{6}$

(b) $\sqrt{ab} \cdot \sqrt{4ab} = \sqrt{4a^2b^2} = 2ab$

(c) $4\sqrt{3a^2} \cdot 2\sqrt{6a^3b} = 8\sqrt{18a^5b} = 24\sqrt{2}a^2\sqrt{a}\sqrt{b} = 24a^2\sqrt{2ab}$

Part 2.

(a) $\dfrac{4x}{\sqrt{7y^2}} = \dfrac{2\sqrt{7x}}{7y}$

(b) $\dfrac{5}{\sqrt[3]{4x^2}} = \dfrac{5\sqrt[3]{2x}}{2x}$

(c) $\dfrac{3\sqrt[3]{xy}}{\sqrt[3]{16x^2y}} = \dfrac{3\sqrt[3]{4x^2}}{4x}$

(d) $\dfrac{\sqrt[4]{2x}}{\sqrt[4]{8x^3}} = \dfrac{\sqrt[4]{4x^2}}{2x} = \dfrac{\sqrt{2x}}{2x}$

10

(a) $5\sqrt{3} + 3\sqrt{2}$

(b) $4\sqrt{2} + 2\sqrt{2} = 6\sqrt{2}$

$\boxed{11}$

(a) $5\sqrt[3]{3}+2\sqrt{3}$

(b) $\sqrt[6]{8}+\sqrt{8}=\sqrt{2}+2\sqrt{2}=3\sqrt{2}$

$\boxed{12}$

(a) $\sqrt{28}-\sqrt{7}=2\sqrt{7}-\sqrt{7}=\sqrt{7}$

(b) $\sqrt{14}+\sqrt{35}=(\sqrt{2}+\sqrt{5})\sqrt{7}$

$\boxed{13}$

(a)

$$\frac{\sqrt{2}-\sqrt{5}}{\sqrt{3}-\sqrt{5}}=\frac{(\sqrt{2}-\sqrt{5})(\sqrt{3}+\sqrt{5})}{(\sqrt{3}-\sqrt{5})(\sqrt{3}+\sqrt{5}}$$
$$=\frac{5+\sqrt{15}-\sqrt{6}-\sqrt{10}}{2}$$

(b)

$$\frac{4\sqrt{x}-\sqrt{3}}{2-\sqrt{7}}=\frac{(4\sqrt{x}-\sqrt{3})(2+\sqrt{7})}{(2-\sqrt{7})(2+\sqrt{7})}$$
$$=\frac{8\sqrt{x}+4\sqrt{7x}-2\sqrt{3}-\sqrt{21}}{-3}$$
$$=\frac{2\sqrt{3}+\sqrt{21}-8\sqrt{x}-4\sqrt{7x}}{3}$$

$\boxed{14}$

(a) $(\sqrt{3}-\sqrt{x})(\sqrt{3}+\sqrt{x})=3-x$

(b) $(\sqrt{8}+\sqrt{x})(\sqrt{x}-2\sqrt{2})=x-8$

$\boxed{15}$

(a) $x^0=1$

(b) $x^{-1}=\dfrac{1}{x}$

(c) $16^{\frac{3}{2}}=64$

(d) $(x^3y^2)^{-\frac{1}{6}}=\dfrac{1}{x^{\frac{1}{2}}y^{\frac{1}{3}}}=\dfrac{\sqrt{x}\sqrt[3]{y^2}}{xy}$

(e) $x^{\frac{1}{2}-\frac{2}{3}}\cdot y^{\frac{1}{3}-(-\frac{2}{3})}=x^{-\frac{1}{6}}\cdot y=\dfrac{y\cdot\sqrt[6]{x^5}}{x}$

(f) $\dfrac{x^{-4}\cdot y^{-1}}{x^2y^{-3}}=x^{-4-2}\cdot y^{-1-(-3)}=\dfrac{y^2}{x^6}$

$\boxed{16}$

(a) $\left(\dfrac{1500}{1000}\right)^{\frac{1}{3}}-1\approx 0.145$, meaning that the inflation rate is roughly 14.5 percent.

(b) Let $F=2P$. Then, $\left(\dfrac{2P}{P}\right)^{\frac{1}{10}}-1\approx 0.072$, meaning that the inflation rate should be roughly 7.2 percent.

$\boxed{17}$

(a) $\left(\dfrac{x^{-\frac{1}{3}}y}{x^{\frac{2}{3}}y^{-\frac{1}{2}}}\right)^2=\left(\dfrac{1}{x}\times y^{\frac{3}{2}}\right)^2=\dfrac{y^3}{x^2}$

(b) $\left(\dfrac{12x^8}{75y^{10}}\right)^{\frac{1}{2}} = \left(\dfrac{4}{25}\right)^{\frac{1}{2}} \times \left(\dfrac{x^8}{y^{10}}\right)^{\frac{1}{2}} = \dfrac{2}{5} \times \dfrac{x^4}{y^5} = \dfrac{2x^4}{5y^5}$

18

(a)
$$\sqrt[3]{2-x} = 4$$
$$2-x = 4^3$$
$$2-x = 64$$
$$x = -62$$

(b)
$$(1+3x)^{\frac{1}{3}} = -2$$
$$(1+3x) = -8$$
$$3x = -9$$
$$x = -3$$

(c)
$$3x^{\frac{3}{2}} - 1 = 5$$
$$3x^{\frac{3}{2}} = 6$$
$$x^{\frac{3}{2}} = 2$$
$$x = \sqrt[3]{4}$$

(d)
$$\sqrt{x} - \sqrt{x-1} = 1$$
$$\sqrt{x} = 1 + \sqrt{x-1}$$
$$x = 1 + 2\sqrt{x-1} + (x-1)$$
$$0 = 2\sqrt{x-1}$$
$$x = 1$$

19

(a) $(3x+1)^{\frac{1}{3}} = 27$ implies that $3x+1 = 3^9$. Hence, $x = \dfrac{3^9 - 1}{3}$.

(b) $3x^{\frac{4}{3}} + 4 = 52$ implies that $x^{\frac{4}{3}} = 16$. Hence, $x = 8$.

(c) $(x-1)^{\frac{2}{3}} = 9$ implies that $x-1 = 27$. Hence, $x = 28$.

20 The area of the square whose perimeter is 2 equals $x = 36$, since $24 = 4\sqrt{x}$.

21

(a)
$$\sqrt{7x-6} = \sqrt{2x+4}$$
$$7x-6 = 2x+4$$
$$5x = 10$$
$$x = 2$$

(b)
$$\sqrt{s+9} - \sqrt{s} = 1$$
$$\sqrt{s+9} = 1 + \sqrt{s}$$
$$s+9 = 1 + 2\sqrt{s} + s$$
$$4 = \sqrt{s}$$
$$16 = s$$

(c)

$$\sqrt{x+2} = x+2$$
$$x+2 = x^2+4x+4$$
$$0 = x^2+3x+2$$
$$0 = (x+2)(x+1)$$
$$x = -1, -2$$

(d)

$$\sqrt{x} = x-6$$
$$x = x^2-12x+36$$
$$0 = x^2-13x+36$$
$$0 = (x-4)(x-9)$$

The only answer is $x = 9$.

(e)

$$\sqrt{x+2}+18 = x$$
$$\sqrt{x+2} = x-18$$
$$x+2 = x^2-36x+324$$
$$0 = x^2-37x+322$$
$$0 = (x-14)(x-23)$$

Here, the only possible answer is $x = 23$.

(f)

$$\sqrt{3-x} = 2x-5$$
$$3-x = 4x^2-20x+25$$
$$0 = 4x^2-19x+22$$
$$0 = (4x-11)(x-2)$$

The only possible answer is $x = \dfrac{11}{4}$.

22

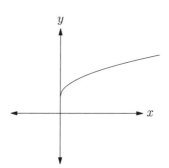

23

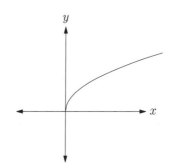

24

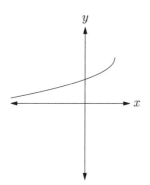

25

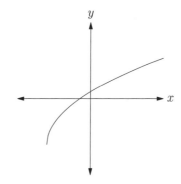

1

(a) Since it is a direct variation, $y = kx$ for $(x,y) = (2,6)$. Hence, $k = 3$. Therefore, $y = 3x$. Substituting $(3,m)$ into the equation, we get $m = 9$.

(b) The answer for (b) is similar to (a). Here, the constant of variation is equal to -2. Hence, $t = (-1)(-2) = 2$.

(c) Let $y = kx$ for some constant of variation k. Then, $y = \dfrac{3}{4}x$. Therefore, $-6 = \dfrac{3}{4}x$. Thus, $x = -8$.

2

(a) Since $k = \dfrac{4}{7}$, we get $y = \dfrac{4}{7}x$. Hence, $10 = \dfrac{4}{7}x$. Thus, $x = \dfrac{35}{2}$.

(b) Since $k = -4$, we get $y = -4x$. Thus, $y = -4(4) = -16$.

3

(a) $2 \times 6 = 3 \times k$, so $k = 4$.

(b) $3 \times -6 = -1 \times x$, so $x = 18$.

(c) $4 \times 3 = x \times -6$, so $x = -2$.

4

(a) Since $y = \dfrac{k}{x^2}$, we get $k = 1600$. Hence, $y = 16$ when $x = 10$.

(b) Since $x = kyz$, we get $k = -\dfrac{1}{2}$. Therefore, $x = -50$ when $y = 5$ and $z = 20$.

5

(a)   (b)

6

(a)

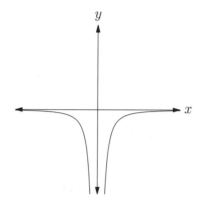

(b)

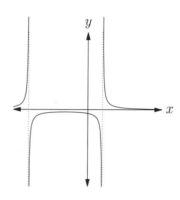

(a) $x = 3$ (b) $x = 1$ (c) $x = -1$

8

(a)

$-1 < x \leq 2$ by caseworks. First, let's find out the critical values that make the fraction either 0 or undefined. The values are -1 and 2. Draw these values on the real number line and find out that there are three intervals to check for signs. If $x < -1$, then the fraction is always positive. If $-1 < x < 2$, the fraction is negative. If $2 < x$, the fraction is positive. At $x = -1$, it is undefined. However, at $x = 2$, the fraction becomes 0. Hence, the right endpoint is included in the solution whereas the left endpoint is not.

(b)

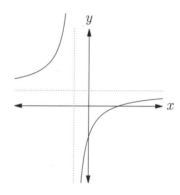

As one can check from the graph, the portion of the graph that is below or equal to the x-axis can be explained if $-1 < x \leq 2$.

9

Let the work rate of the first pipe be P pool per hour. Then, $P = \dfrac{1}{4}$. Let x be the hour spent by the second pipe to fill the pool on its own. Then, the second pipe fills $\dfrac{1}{x}$ pool per hour. Hence,

$$2 \times \left(\frac{1}{4} + \frac{1}{x} \right) = 1$$

Hence, $x = 4$ hours.

10

From his office to his house, he spent 30 minutes. On the other hand, from his house to his office, he spent 1 hour. Hence, in order to travel 100 miles in total, he spent 1 hour and 30 minutes. Therefore, the average speed must be $\frac{100}{3/2}(= 200/3)$ miles per hour.

11

$$\frac{1}{x-3} + \frac{1}{x+3} = \frac{4}{x^2-9}$$
$$\frac{2x}{x^2-9} = \frac{4}{x^2-9}$$
$$2x = 4$$
$$x = 2$$

12

$$\frac{1}{2x} - \frac{1}{3(x+5)} = \frac{1}{x}$$
$$\frac{1}{2x} - \frac{2}{2x} = \frac{1}{3(x+5)}$$
$$-\frac{1}{2x} = \frac{1}{3(x+5)}$$
$$-2x = 3(x+5)$$
$$x = -3$$

1

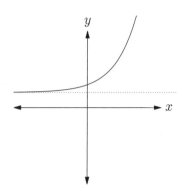

2

(a) The domain is the set of all real numbers, i.e., $\mathbb{R}$.

(b) The range equals $(-2, \infty)$.

(c) The horizontal asymptote is $y = -2$.

3

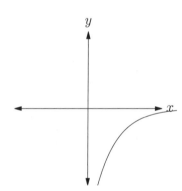

Although the figure is not drawn to include the y-intercept, there exists the y-intercept of -8 at $x = 0$. The domain is $\mathbb{R}$, the range is $(-\infty, 0)$, and the horizontal asymptote is $y = 0$.

4

(a)

$$4^{2x+2} = 8$$
$$2^{4x+4} = 2^3$$
$$4x + 4 = 3$$
$$4x = -1$$
$$x = -\frac{1}{4}$$

(b)

$$3^{x+2} = \frac{1}{27}$$
$$3^{x+2} = 3^{-3}$$
$$x + 2 = -3$$
$$x = -5$$

(c)

$$3(9)^{x+3} = \frac{1}{27^x}$$
$$3^1 3^{2x+6} = 3^{-3x}$$
$$3^{2x+7} = 3^{-3x}$$
$$2x + 7 = -3$$
$$5x = -7$$
$$x = -\frac{7}{5}$$

5

(a)

$$9(3^x) = 27^{1-x}$$
$$3^2 \cdot 3^x = (3^3)^{1-x}$$
$$3^{2+x} = 3^{3(1-x)}$$
$$3^{2+x} = 3^{3-3x}$$
$$2 + x = 3 - 3x$$
$$4x = 1$$
$$x = \frac{1}{4}$$

(b)

$$16^{x-1} = \frac{\sqrt{2}}{8^{-x}}$$
$$2^{4x-4} = 2^{\frac{1}{2}+3x}$$
$$4x - 4 = \frac{1}{2} + 3x$$
$$x = \frac{9}{2}$$

(c)

$$\frac{2^{x+2}}{16} = 4^{2x-1}$$
$$2^{x+2-4} = 2^{4x-2}$$
$$x - 2 = 4x - 2$$
$$x = 0$$

(d)

$$9^{\sqrt{x+1}} = \frac{1}{3^{7-x}}$$
$$3^{2\sqrt{x+1}} = 3^{x-7}$$
$$2\sqrt{x+1} = x - 7$$
$$4(x+1) = x^2 - 14x + 49$$
$$0 = x^2 - 18x + 45$$
$$0 = (x-3)(x-15)$$

However, $x = 3$ is extraneous. Hence, $x = 15$.

(e)

$$4^{x^2+2x} = \frac{1}{64}$$
$$4^{x^2+2x} = 4^{-3}$$
$$x^2 + 2x = -3$$
$$x^2 + 2x + 3 = 0$$

There is no real solution.

6

(a)

$$3^{2x} - 4(3^x) + 3 = 0$$
$$(3^x - 3)(3^x - 1) = 0$$
$$x = 1, 0$$

(b)

$$2^{2x} - 2^{x+1} - 8 = 0$$
$$(2^x - 4)(2^x + 2) = 0$$
$$x = 2$$

(c)

$$5^{2x} + 25 = 26(5^x)$$
$$5^{2x} - 26(5^x) + 25 = 0$$
$$(5^x - 25)(5^x - 1) = 0$$
$$x = 2, 0$$

(d)

$$3^{1+x} = \frac{9}{3^x} + 26$$
$$3(3^x)^2 - 26(3^x) - 9 = 0$$
$$3^x = -1/3, 9$$
$$x = 2$$

(e)

$$2^{2x} + 2^{x+2} = 8$$
$$(2^x)^2 + 4(2^x) - 8 = 0$$
$$2^x = -2 + 2\sqrt{3}$$
$$x = \log_2(-2 + 2\sqrt{3})$$

$\boxed{7}$ According to the compound interest rate formula, we get $1500(1.04)^4 \approx 1754.78 \approx 1755$ dollars.

$\boxed{8}$ According to the population growth model, we get $130,000(1.02)^{15} \approx 174,963$ number of people after 15 years.

$\boxed{9}$ Let's find out the percent of decrease by $\dfrac{72000 - 56000}{72000} \approx 22\%$. Hence, the exponential model that models this would be given by

$$72000(1 - \frac{2}{9})^3 \approx 33876.5432 \approx 33877(\text{ in dollars })$$

$\boxed{10}$

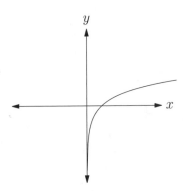

$\boxed{11}$

(a) $\log_2(32) = 5$

(b) $\log_2\left(\dfrac{1}{2}\right) = -1$

(c) $\log_{(x+1)}(4) = 2$

(d) $\log_a(b) = -m$

12

(a) $2^4 = 16$

(b) $\left(\dfrac{3}{2}\right)^2 = \dfrac{9}{4}$

(c) $(27)^{\frac{4}{3}} = 81$

13

(a)

$$\log_2(8) = x$$
$$2^x = 8$$
$$x = 3$$

(b)

$$\log_3(27) = x$$
$$3^x = 27$$
$$x = 3$$

(c)

$$\log_3(3) = x$$
$$3^x = 3$$
$$x = 1$$

(d)

$$\log_4(1) = x$$
$$4^x = 1$$
$$x = 0$$

14

(a)

$$\log_{\sqrt{2}}(x) = 2$$
$$(\sqrt{2})^2 = x$$
$$2 = x$$

(b)

$$\log_x(\sqrt{2}) = 4$$
$$x^4 = \sqrt{2}$$
$$x = 2^{\frac{1}{8}}$$
$$x = \sqrt[8]{2}$$

(c)

$$\log_{(x-1)}\sqrt{5} = \dfrac{1}{2}$$
$$(x-1)^{\frac{1}{2}} = \sqrt{5}$$
$$x - 1 = 5$$
$$x = 6$$

15

(a)

$$\log_3(9 \times 27)$$
$$= \log_3(3^5)$$
$$= 5$$

(b)

$$\log_3(8) - \log_3(3)$$
$$= \log_3\left(\dfrac{8}{3}\right)$$

(c)

$$\log_2(3^2) + \log_2(9^3)$$
$$= \log_2(9^4)$$
$$= \log_2(6561)$$

16

(a) $3^{\log_3(x)} = x$

(b) $e^{\ln(4)} = e^{\log_e(4)} = 4$

(c) $2^{2\log_2(4)} = 2^{\log_2(16)} = 16$

17

(a)

$$\frac{\log(3)}{\log(2)} \times \frac{\log(5)}{\log(3)} \times \frac{\log(4)}{\log(5)}$$
$$= \frac{\log(4)}{\log(2)}$$
$$= \log_2(4)$$
$$= 2$$

(b)

$$e^{\ln(3)} \times 5^{2\log_{\sqrt{5}}(4)}$$
$$= 3 \times 5^{\log_{\sqrt{5}}(16)}$$
$$= 3 \times (\sqrt{5})^{2\log_{\sqrt{5}}(16)}$$
$$= 3 \times 256$$
$$= 768$$

(c)

$$\log_{125}(5) + \log_{27}(9) = \frac{1}{3} + \frac{2}{3}$$
$$= 1$$

(d)

$$\frac{\log_2(25)}{\log_2(5)} \times \frac{\log(16)}{\log(4)} = 2 \times 2$$
$$= 4$$

18

(a) $x + y$

(b) $x + y - 1$

(c) $\frac{1}{x}$

(d) $\frac{x+y}{x}$

(e) x

(f) $\frac{1+x}{y}$

19

(a)

$$3^x - 4 = 7$$
$$3^x = 11$$
$$x = \log_3(11)$$

(b)

$$5^{2x} - 3 = 4$$
$$5^{2x} = 7$$
$$2x = \log_5(7)$$
$$x = 1/2 \cdot \log_5(7)$$

(c)

$$100e^{2x} = 200$$
$$e^{2x} = 2$$
$$2x = \log_e(2)$$
$$2x = \ln(2)$$
$$x = 1/2 \cdot \ln(2)$$

20

(a)

$$2^{2x} - 4(2x) = 5$$
$$(2^x - 5)(2^x + 1) = 0$$
$$2^x = 5$$
$$x = \log_2(5)$$

(b)

$$3^x - 3(3^{-x}) - 2 = 0$$
$$(3^x - 3)(3^x + 1) = 0$$
$$3^x = 3$$
$$x = \log_3(3)$$
$$x = 1$$

21

(a)

$$3^{2x} = 10^{1-x}$$
$$2x\ln(3) = (1-x)\ln(10)$$
$$(2\ln(3) + \ln(10))x = \ln(10)$$
$$x = \frac{2\ln(3) + \ln(10)}{\ln(10)}$$

(b)

$$4^{x-1} = 3^{2+x}$$
$$(x-1)\ln(4) = (2+x)\ln(3)$$
$$(\ln(4) - \ln(3))x = \ln(4) + 2\ln(3)$$
$$x = \frac{\ln(4) + 2\ln(3)}{\ln(4) - \ln(3)}$$

(c)

$$\frac{m^{2x-1}}{n^x} = p^{2-x}$$
$$(2x-1)\log(m) - x\log(n) = (2-x)\log(p)$$
$$(2\log(m) - \log(n) + \log(p))x = 2\log(p) + \log(m)$$
$$x = \frac{2\log(p) + \log(m)}{2\log(m) - \log(n) + \log(p)}$$

22

(a)

$$\log_2(3x) = 8$$
$$2^8 = 3x$$
$$\frac{256}{3} = x$$

(b)

$$\log(2x+3) = -2$$
$$10^{-2} = 2x+3$$
$$\frac{1}{100} - 3 = 2x$$
$$-\frac{299}{200} = x$$

(c)

$$\log_3(1+x^2) = 27$$
$$3^{27} = 1 + x^2$$
$$3^{27} - 1 = x^2$$
$$\pm\sqrt{3^{27} - 1} = x$$

23

(a)

$$\log_3(\log_2(x)) = 1$$
$$\log_2(x) = 3$$
$$2^3 = x$$
$$8 = x$$

(b)

$$\log_4(\log_{27}(x-1)) = \frac{1}{2}$$
$$\log_{27}(x-1) = 2$$
$$27^2 = x - 1$$
$$27^2 + 1 = x$$
$$730 = x$$

$\boxed{24}$

(a)

$$\log_5(5x+40) - \log_5(x+2) = 2$$
$$\log_5\left(\frac{5x+40}{x+2}\right) = 2$$
$$25 = \frac{5x+40}{x+2}$$
$$25x+50 = 5x+40$$
$$20x = -10$$
$$x = -\frac{1}{2}$$

(b)

$$\log_3(3-2x) + \log_3(x+1) = 1$$
$$(3-2x)(x+1) = 3$$
$$-2x^2 + x = 0$$
$$x = 0, \frac{1}{2}$$

(c)

$$\ln(x-9) + \ln(x) = \ln(10)$$
$$\ln(x^2 - 9x) = \ln(10)$$
$$x^2 - 9x - 10 = 0$$
$$x = 10, -1$$

Here, $x = -1$ is extraneous. Thus, $x = 10$.

(d)

$$\log_2(x-1) - \log_4(x+2) = 1$$
$$\log_2(x-1) - \log_2\sqrt{x+2} = 1$$
$$\log_2\left(\frac{x-1}{\sqrt{x+2}}\right) = 1$$
$$x-1 = 2\sqrt{x+2}$$
$$x^2 - 6x - 7 = 0$$
$$x = 7, -1$$

Here, $x = -1$ is extraneous. Thus, $x = 7$.

1

(a) $a_n = \dfrac{2n+1}{2^{n+1}}$

(b) $b_k = b_{k-1} + b_{k-2}$ for $k \geq 3$, where $b_1 = 1$, $b_2 = 1$.

2 Since $a_n = 4 + (n-1)5 = 5n - 1$, so $a_{20} = 99$.

3 Since $a_3 = a_1 + 2d = 13$ and $a_8 = a_1 + 7d = 33$, so $(a_1, d) = (5, 4)$. Hence, $a_{27} = a_1 + 26d = 109$.

4

(a) The annual salary must be $a_5 = a_1 + 4d = 50,000 + 4(1200) = 54,800$.

(b) The total salary he earns in ten years must be $\dfrac{10(2(50,000) + 9(1,200))}{2} = 554,000$ dollars.

5 Since $a_n = 3 \times 2^{n-1} = 3072$, we get $n = 11$.

6 Since we have three consecutive terms of geometric sequence, $(5k-3)^2 = (k+3)(7k+3)$, $25k^2 - 30k + 9 = (7k+3)(k+3) = 7k^2 + 24k + 9$, so $k = 0$ or $k = 3$.

7 $1 + \dfrac{1}{2} + \dfrac{1}{4} + \cdots = \dfrac{1}{1 - 1/2} = 2.$

8 $3 + 1 + \dfrac{1}{3} + \cdots = \dfrac{3}{1 - 1/3} = \dfrac{9}{2}.$

9 Since $10,000(1 - 0.02)^t = P(t)$, the answer must be (B).

10 $1,000(1.03)^3 \approx 1,060.9$, according to the formula.

11 The arithmetic mean of 2 and 18 is $\dfrac{2+18}{2} = 10$. On the other hand, the geometric mean of 2 and 18 equals $\sqrt{2 \times 18} = 6$.

12 Since three arithmetic sequential terms are consecutive, we use the idea of arithmetic mean, i.e.,

$$\frac{(2x+3) + (10x - 15)}{2} = 5x - 2$$
$$12x - 12 = 10x - 4$$
$$2x = 8$$
$$x = 4$$

$\boxed{13}$

(a) Substituting 1, 2, and 3, we get $a+b+c=0$, $4a+2b+c=2$, and $9a+3b+c=6$. Solving the system of linear equations, we get $(a,b,c)=(1,-1,0)$. Hence, $x_n = n^2 - n$.

(b) Substituting 1, 2, and 3, we get $a+b+c=2$, $4a+2b+c=5$, and $9a+3b+c=10$. Hence, we get $x_n = n^2+1$.

$\boxed{14}$

(a) Since there is only one layer of difference, a_n must be linear. Also, the last difference is 4, so $a_n = 4n+b$. Substituting $n=1$, we get $b=1$. Hence, $a_n = 4n+1$.

(b) Since there are two layers of differences, a_n must be quadratic. Since the difference in the last layer is 2, we get $a_n = n^2+bn+c$. Substituting $n=1$ and $n=2$ results in $(b,c)=(3,2)$. Hence, $a_n = n^2+3n+2$.

(c) Since there are three layers of differences, a_n must be cubic. Since the difference in the last layer is 3, we get $a_n = 1/2 \cdot n^3 + bn^2 + cn + d$. Substituting $n=1,2,3$ results in $a_n = 1/2 \cdot n^3 + 3/2 \cdot n$.

$\boxed{15}$

(a)

$$\sum_{k=1}^{20}(3k+4) = 3\sum_{k=1}^{20}k + \sum_{k=1}^{20}4$$
$$= 3\times\frac{20\times 21}{2} + 4\times 20$$
$$= 630+80$$
$$= 710$$

(b)

$$\sum_{k=1}^{15}(2k^2+4) = 2\sum_{k=1}^{15}k^2 + \sum_{k=1}^{15}4$$
$$= 2\times\frac{15\times 16\times 31}{6} + 4\times 15$$
$$= 2480+60$$
$$= 2540$$

$\boxed{16}$

(a)

$$\sum_{k=11}^{25}k^3 = \sum_{k=1}^{25}k^3 - \sum_{k=1}^{10}k^3$$
$$= \left(\frac{25\times 26}{2}\right)^2 - \left(\frac{10\times 11}{2}\right)^2$$
$$= (25\times 13)^2 - (5\times 11)^2$$
$$= 102600$$

(b)

$$\sum_{k=6}^{18}(k^2+3k+2) = \sum_{k=6}^{18}k^2 + \sum_{k=6}^{18}3k + \sum_{k=6}^{18}2$$

$$= \left(\sum_{k=1}^{18}k^2 - \sum_{k=1}^{5}k^2\right) + \left(\sum_{k=1}^{18}3k - \sum_{k=1}^{5}3k\right) + \left(\sum_{k=1}^{18}2 - \sum_{k=1}^{5}2\right)$$

$$= \left(\frac{18\times19\times37}{6} - \frac{5\times6\times11}{6}\right) + 3\left(\frac{18\times19}{2} - \frac{5\times6}{2}\right) + 2(18-5)$$

$$= 2548$$

(c)

$$\sum_{k=1}^{8}2^{-k} = \frac{1}{2} + \frac{1}{4} + \frac{1}{8} + \cdots + \frac{1}{256}$$

$$= \frac{1}{256}\left(1+2+2^2+2^3+2^4+\cdots+2^7\right)$$

$$= \frac{255}{256}$$

17 Let's use $a_n = S_n - S_{n-1}$. Hence, $a_n = n^2 - (n-1)^2 = 2n-1$.

Answers in this topic are mostly not simplified for readers so as to focus on the procedures of retrieving the correct answers.

1 There are 119 integers between 56 and 174, inclusive.

2 There are 216 integers. We use 1-to-1 correspondence to find out that the index of each number in the sequence is half the number itself.

3

(a) $26 \times 26 \times 26 \times 26$ four-letter words.

(b) There are $4 \times 3 \times 2$ different apparel styles.

(c) There are $4 \times 4 \times 4$ possible outcomes.

4

(a) Let's case-enumerate. If the last digit is 0, then there are $9 \times 8 \times 7 \times 6$ possible even digit numbers. If the last digit is 2, then there are $8 \times 8 \times 7 \times 6$ possible even digit numbers. If the last digit is 4, then there are $8 \times 8 \times 7 \times 6$ possible even digit numbers. If the last digit is 6, then there are $8 \times 8 \times 7 \times 6$ possible even digit numbers. If the last digit is 8, there are $8 \times 8 \times 7 \times 6$ possible even digit numbers. Hence, there are $9 \times 8 \times 7 \times 6 + 4 \times 8 \times 8 \times 7 \times 6$ possible even 5-digit numbers.

(b) There are $5 \times 21 \times 21 \times 5$ number of four-letter words satisfying the given condition.

(c) Let the first digit be 4. Then, there are $3 \times 3 \times 2$ possible ways of producing the numbers. On the other hand, let the first digit be 8. Then, there are $4 \times 3 \times 2$ possible ways of producing the numbers in this case. Hence, there are $18 + 24$ possible four-digit numbers satisfying the given condition.

5

(a) There are $5 \times 5 \times 5$ ways Jimmy can give Skittles to three kids.

(b) The winning team has three possible ways of scoring the goals. It could score none, 1 goal or two goals. On the other hand, the losing team has only one option of scoring no goal. Hence, there are 3×1 possibilities.

6

(a) $1 \times 5 \times 5 \times 10 \times 10$ odd numbers with third digit 5 are between 20000 and 69999.

(b) Let's case enumerate. If there is no N used, then there are $5 \times 4 \times 3 \times 2$ number of four-letter words. If there is one N used, then there are $4 \times 5 \times 4 \times 3$ number of four-letter words. If there are two N's used in four-letter words, then there are $6 \times 5 \times 4$ number of four-letter words. Hence, there are 480 number of four-letter words.

$\boxed{7}$

(a) There are 8 positive divisors of 24.

(b) There are 7×7 positive divisors of $1,000,000$.

(c) There are 4×5 positive divisors of $n = p^3 q^4$.

$\boxed{8}$

(a) There are $3 \times 2 \times 10 \times 10 \times 9$ five-letter word satisfying the condition.

(b) Let's case-enumerate. If the last digit is 0, then there are $1 \times 6 \times 5 \times 4$ numbers formed. Similarly, if the last digit is 5, then there are $1 \times 5 \times 5 \times 4$ numbers formed. Hence, in total, 220 numbers formed.

$\boxed{9}$

(a) $6 \times 5 \times 4 \times 3$ (b) 10×9 (c) $4 \times 3 \times 2 \times 1$

$\boxed{10}$

(a)

$$2 \times {}_nP_2 = 3 \times {}_{n-1}P_2$$
$$2n(n-1) = 3(n-1)(n-2)$$
$$(n-1)(2n-3n+6) = 0$$
$$n = 1, 6$$

Here, $n = 6$.

(b)

$$_{n+1}P_2 = 50 \times {}_nP_1$$
$$(n+1)n = 50n$$
$$(n+1-50)n = 0$$
$$(n-49)n = 0$$

Here, $n = 49$.

$\boxed{11}$

(a) $6! \times 2 = 720$ (b) $7! - 6! \times 2 = 5 \times 6!$

$\boxed{12}$

(a) $4! \times 2! = 48$ (b) $3! \times 3! = 36$

$\boxed{13}$ Circular permutation of 5 students equals $1 \times 4!$.

$\boxed{14}$ It does not matter whom we choose to be our observer. Let A be the first one to be seated. Then, B has two possible seats. Hence, there are $1 \times 2 \times 4! = 48$ ways.

$\boxed{15}$ Three couples can be seated around a circular table such that each couple should sit together, in $1 \times 2 \times (1 \times 1 \times 2 \times 1 + 1 \times 1 \times 2 \times 1 + 1 \times 1 \times 2 \times 1 + 1 \times 1 \times 2 \times 1) = 16$ ways.

$\boxed{16}$ Let the first adult be seated. Then, there are $2!$ possible seats for two adults. Also, there are $3!$ possible ways for three children to be seated. Hence, 12 ways are there.

$\boxed{17}$ By case-enumeration, we get $2(12 + 15 + 20) = 94$ ways.

18 Let's case-enumerate. If the last digit is 1, then there are $1 \times 1 \times 4 \times 7 \times 6$ odd 5-digit numbers satisfying the given condition. If the last digit is 3, there are $1 \times 1 \times 3 \times 7 \times 6$ odd 5-digit numbers. If the last digit is 7, there are $1 \times 1 \times 4 \times 7 \times 6$ odd 5-digit numbers satisfying the given condition. If the last digit is 9, there are $1 \times 1 \times 4 \times 7 \times 6$ odd 5-digit numbers satisfying the given condition, as well. Hence, there are $168 \times 3 + 126 = 630$ odd 5-digit numbers satisfying the condition.

19 If three babies are put in two different playpens, then there are 8 possible ways to place the babies in the playpens. On the other hand, if three babies are put in two identical-looking playpens, there are 4 possible ways to place them in playpens.

20 This is an application of partition of natural numbers. In other words, we are solving for $x+y = 3$. Then, $(x,y) = (3,0),(2,1),(1,2),(0,3)$. Instead of stopping at $(3,0)$ and $(2,1)$, we find the other two possibilities because the babies are distinguishable. There are four possible ways of giving three identical-looking rattles to two babies.

21 The team must have $WWW, LWWW, WLWW$, or $WWLW$. Hence, there are four possible outcomes for the team to go up to the finals.

22 There are 8 possible ways, i.e., $OOOO, FOOO, OFOO, OOFO, OOOF, OFOF, FOFO$, and $FOOF$.

23

(a) $_5C_2 = \dfrac{5 \times 4}{2 \times 1}$ (b) $_5C_5 = 1$ (c) $_5C_0 = 1$

24 $_nC_2 = 55 = \dfrac{n(n-1)}{2} = 55$, so $n(n-1) = 110$. Hence, $n = 11$.

25 Choose 2 colors out of 5 colors. In other words, she has $_5C_2 = \dfrac{5 \times 4}{2 \times 1} = 10$ ways to choose two colors out of five.

26 If the values on the die are equal, then we have 6 possible cases. If the values on the die are distinct, then there are $_6C_2 = 15$ possible cases. Hence, there are 21 possible cases.

27

(a) $_{100}C_{99} = 100$ (b) $_{n+1}C_{n-1} = \dfrac{(n+1)n}{2}$

28 Solving $_nC_3 = {}_nC_5$, we get $n = 3+5$. Hence, the polygon must be octagon.

29 RAMANUJAN can be rearranged in $\dfrac{9!}{3!2!}$ ways, including itself. Similarly, MINIMIZATION can be rearranged in $\dfrac{12!}{2!2!4!}$ ways, including itself.

30 Just like question 29, there are $\dfrac{9!}{3!2!}$ number of arrangements of $ABCDEFAAB$.

31

(a) $32x^5 + 240x^4y + 720x^3y^2$

(b) $x^4 + 4x^2 + 6$

(c) $3^{10}x^{10} + 10 \times 3^9 x^9 y + 45 \times 3^8 x^8 y^2$

32 The constant term of the expansion can be found by manipulating
$_6C_k(x^2)^k\left(-\dfrac{2}{x}\right)^{6-k} = {_6}C_2(-2)^{6-k}x^{3k-6}$. We need to make the exponent of x as 0. Hence, $k = 2$.
Therefore, the constant term is 240.

33 Using the generating function, we are looking for the coefficient of x^8 from
$(1+x)^{10}(1+x+x^2+x^3)$, where each exponent is the dollar amount used by one person. Hence, the answer must be $_{10}C_2 + {_{10}}C_3 + {_{10}}C_4 + {_{10}}C_5$.

1 The sample space consists of all possible cases. Hence, it must be $\{HHH, HHT, HTH, THH, HTT, THT, TTH, TTT\}$.

2 Out of 36 possible cases, there are 5 possible cases to produce the sum of 6. There are 4 possible cases to produce the sum of 5. There are 3 possible cases to produce the sum of 4. There are 2 possible cases to produce the sum of 3. There is 1 possible case to produce the sum of 2. Hence, the probability must be $\dfrac{15}{36} = \dfrac{5}{12}$.

3

(a) $P(A') = \dfrac{1}{2}$

(b) $P(A \cup B) = \dfrac{3}{4}$

(c) $P(A|B) = \dfrac{1}{3}$

4

(a) $P(A \cap B) = \dfrac{1}{18}$

(b) $P(A|B) = \dfrac{1}{5}$

(c) A and B are independent if $P(A \cap B) = P(A) \cdot P(B)$. This is not true, so they are not independent.

(d) A and B are mutually exclusive if $P(A \cap B) = 0$. This is not true, so they are not mutually exclusive.

5

(a) $P(A) = \dfrac{2}{3}$

(b) $P(B) = \dfrac{1}{3}$

(c) $P(B|A) = \dfrac{3}{10}$

(d) $P(B|A') = \dfrac{2}{5}$

(e) $P(B'|A') = \dfrac{3}{5}$

(f) $P(A \cap B | A \cup B) = \dfrac{1}{4}$

6

	1	2	3	4	5	6
1	2	3	4	5	6	7
2	3	4	5	6	7	8
3	4	5	6	7	8	9
4	5	6	7	8	9	10
5	6	7	8	9	10	11
6	7	8	9	10	11	12

(a) $\dfrac{5+4+3+2+1}{36} = \dfrac{5}{12}$

(b) $\dfrac{2+5+4+1}{36} = \dfrac{1}{3}$

(c) $\dfrac{3+2+1}{36} = \dfrac{1}{6}$

(d) $\dfrac{1+2+3+4}{36} = \dfrac{5}{18}$

7

(a) $\dfrac{10}{13} \times \dfrac{9}{12} \times \dfrac{8}{11} \times \dfrac{7}{10} = \dfrac{42}{143}$

(b) $\dfrac{6}{13} \times \dfrac{5}{12} \times \dfrac{4}{11} \times \dfrac{3}{10} = \dfrac{3}{143}$

(c) $\dfrac{4}{13} \times \dfrac{9}{12} \times \dfrac{8}{11} \times \dfrac{7}{10} \times 4 = \dfrac{336}{715}$

(d) $\dfrac{6 \cdot 5 \cdot 4 \cdot 2 + 4 \cdot 3 \cdot 2 \cdot 1}{13 \cdot 12 \cdot 11 \cdot 10} = \dfrac{1}{65}$

8

(a) $1 - \dfrac{6}{10} \times \dfrac{5}{9} \times \dfrac{4}{8} = \dfrac{5}{6}$

(b) $\dfrac{6}{10} \times \dfrac{5}{9} \times \dfrac{4}{8} + \dfrac{4}{10} \times \dfrac{6}{9} \times \dfrac{5}{8} \times 3 = \dfrac{2}{3}$

(c) $\dfrac{6}{10} \times \dfrac{5}{9} \times \dfrac{4}{8} = \dfrac{1}{6}$

(d) $\dfrac{4}{10} \times \dfrac{3}{9} \times \dfrac{2}{8} \times 1 = \dfrac{1}{30}$

(e) $\dfrac{4}{10} \times \dfrac{3}{9} \times \dfrac{6}{8} \times 3 + \dfrac{4}{10} \times \dfrac{3}{9} \times \dfrac{2}{8} \times 1 = \dfrac{1}{3}$

9

(a) $1 - \dfrac{48}{52} \times \dfrac{47}{51} \times \dfrac{46}{50} \times \dfrac{45}{49} = \dfrac{15229}{54145}$

(b) $\dfrac{13}{52} \times \dfrac{12}{51} \times \dfrac{11}{50} \times \dfrac{10}{49} = \dfrac{11}{4165}$

(c) $\dfrac{39}{52} \times \dfrac{38}{51} \times \dfrac{37}{50} \times \dfrac{36}{49} + \dfrac{13}{52} \times \dfrac{39}{51} \times \dfrac{38}{50} \times \dfrac{37}{49} \times 4 = \dfrac{15466}{20825}$

(d) $\dfrac{4}{52} \times \dfrac{3}{51} \times \dfrac{2}{50} \times \dfrac{1}{49} \times 13 = \dfrac{1}{20825}$

10 First, it should not rain. Hence, the probability that it will not rain equals $\dfrac{45}{100}$. When it does not rain, the probability of going to the beach equals $\dfrac{70}{100}$. We multiply the two probabilities together, i.e., $\dfrac{45}{100} \times \dfrac{70}{100} = 0.315$.

11

(a) $\dfrac{6!}{2!4!} \cdot \left(\dfrac{2}{5}\right)^2 \cdot \left(\dfrac{3}{5}\right)^4 = \dfrac{972}{3125}$

(b) $\left(\dfrac{3}{5}\right)^6 + \dfrac{6!}{1!5!} \cdot \left(\dfrac{2}{5}\right) \cdot \left(\dfrac{3}{5}\right)^5 = \dfrac{729}{3125}$

(c) $1 - \left(\left(\dfrac{3}{5} \right)^6 + \dfrac{6!}{1!5!} \cdot \left(\dfrac{2}{5} \right) \cdot \left(\dfrac{3}{5} \right)^5 \right) = \dfrac{2396}{3125}$

12

(a) $\dfrac{4!}{3!1!} \cdot \left(\dfrac{2}{5} \right)^3 \cdot \left(\dfrac{3}{5} \right) = \dfrac{96}{625}$

(c) $\left(\dfrac{2}{5} \right)^5 + \dfrac{5!}{4!1!} \cdot \left(\dfrac{2}{5} \right)^4 \cdot \left(\dfrac{3}{5} \right) + \dfrac{5!}{3!2!} \cdot \left(\dfrac{2}{5} \right)^3 \cdot \left(\dfrac{3}{5} \right)^2 = \dfrac{992}{3125}$

13

(a) $\dfrac{4}{10} \times \dfrac{3}{9} \times \dfrac{2}{8} \times 1 = \dfrac{1}{30}$

(b) $\dfrac{4}{10} \times \dfrac{3}{9} \times \dfrac{6}{8} \times \dfrac{3!}{2!1!} = \dfrac{3}{10}$

(c) $\dfrac{4}{10} \times \dfrac{6}{9} \times \dfrac{5}{8} \times \dfrac{3!}{2!1!} = \dfrac{1}{2}$

(d) $\dfrac{6}{10} \times \dfrac{5}{9} \times \dfrac{4}{8} \times 1 = \dfrac{1}{6}$

14

(a) $\dfrac{1}{6} \times \dfrac{1}{6} \times \dfrac{1}{6} \times 6 = \dfrac{1}{36}$

(b) $1 \times \dfrac{5}{6} \times \dfrac{4}{6} = \dfrac{5}{9}$

15

(a) $4 \times \dfrac{13}{52} \times \dfrac{12}{51} \times \dfrac{11}{50} \times \dfrac{10}{49} = \dfrac{44}{4165}$

(b) $\dfrac{13}{52} \times \dfrac{13}{51} \times \dfrac{13}{50} \times \dfrac{13}{49} \times 4! = \dfrac{2197}{20825}$

16

(a) $\dfrac{4}{52} \times \dfrac{3}{51} \times \dfrac{2}{50} \times \dfrac{4}{49} \times \dfrac{3}{48} = \dfrac{1}{1082900}$

(b) $\dfrac{4}{52} \times \dfrac{3}{51} \times \dfrac{2}{50} \times \dfrac{4}{49} \times \dfrac{3}{48} \times \dfrac{5!}{2!3!} = \dfrac{1}{108290}$

(c) $\dfrac{4}{52} \times \dfrac{3}{51} \times \dfrac{4}{50} \times \dfrac{3}{49} \times \dfrac{4}{48} \times \dfrac{5!}{2!2!1!} = \dfrac{3}{54145}$

17

(a) $\dfrac{4}{9} \times \dfrac{3}{8} \times \dfrac{2}{7} = \dfrac{1}{21}$

(b) $\dfrac{3}{9} \times \dfrac{6}{8} \times \dfrac{5}{7} \times \dfrac{3!}{1!2!} = \dfrac{15}{28}$

(c) $\dfrac{2}{9} \times \dfrac{1}{8} \times \dfrac{7}{7} \times \dfrac{3!}{2!1!} = \dfrac{1}{12}$

18

(a) $\dfrac{4}{9} \times \dfrac{3}{8} \times 1 = \dfrac{1}{6}$

(b) $\dfrac{5}{9} \times \dfrac{4}{8} \times 1 = \dfrac{5}{18}$

(c) $\dfrac{5}{9} \times \dfrac{4}{8} \times \dfrac{2!}{1!1!} = \dfrac{5}{9}$

(d) $1 - \left(\dfrac{5}{9} \times \dfrac{4}{8} \right) = \dfrac{13}{18}$

19 The probability that a student at You-know-what school is left-handed and wearing glasses equals $\dfrac{2}{10} \times \dfrac{4}{13} = \dfrac{4}{65}$.

20

(a) $0.6 \times 0.92 = 0.552 = \dfrac{69}{125}$

(b) $0.4 \times 0.92 = 0.368 = \dfrac{46}{125}$

21

(a) $0.95 \times 0.85 = 0.8075 = \dfrac{323}{400}$

(b) $0.05 \times 0.85 + 0.95 \times 0.15 = 0.185 = \dfrac{37}{200}$

22

(a) $\dfrac{1}{2} \times \dfrac{4}{10} + \dfrac{1}{2} \times \dfrac{7}{10} = \dfrac{11}{20}$

(b) $\dfrac{4/20}{11/20} = \dfrac{4}{11}$

1 The mean value equals

$$E(X) = \frac{2+6+12+16+15}{16} = \frac{51}{16}$$

Similarly, the variance equals

$$Var(X) = \frac{2+12+36+64+75}{16} - \left(\frac{51}{16}\right)^2 = \frac{423}{256}$$

2 Since the sum of probabilities is equal to 1, we get $k+2k+3k+2k+k = 9k = 1$, so $k = \frac{1}{9}$. Hence, we can compute the value of mean and variance.

- Mean : $\dfrac{1+4+9+8+5}{9} = \dfrac{27}{9} = 3$

- Variance : $\dfrac{1+8+27+32+25}{9} - 9 = \dfrac{4}{3}$

3

(a) If $9k \neq 1$, then the probability distribution is false.

(b) Since the probability cannot be less than 0, the probability distribution is false.

4 Since $E(X) = k + 2(0.8 - k) + 0.6 = 1.7$, then $k = 0.5$.

5

(a) $\left(\dfrac{3}{5}\right)^4 \times \left(\dfrac{2}{5}\right)^3 \times \dfrac{7!}{3!4!} = \dfrac{4536}{15625} = \text{binompdf}(7, 0.6, 4)$.

(b) $P(X=0) + P(X=1) + P(X=2) + P(X=3) = \dfrac{7!}{7!0!}\left(\dfrac{2}{5}\right)^7\left(\dfrac{3}{5}\right)^0 + \dfrac{7!}{6!1!}\left(\dfrac{2}{5}\right)^6\left(\dfrac{3}{5}\right)^1 +$

$\dfrac{7!}{5!2!}\left(\dfrac{2}{5}\right)^5\left(\dfrac{3}{5}\right)^2 + \dfrac{7!}{3!4!}\left(\dfrac{2}{5}\right)^4\left(\dfrac{3}{5}\right)^3 = \text{binomcdf}(7, 0.6, 3) = 0.289792$.

(c) The expected number of rainy days must be $\dfrac{21}{5} = 4.2$ days. Its variance must be $7 \times \dfrac{3}{5} \times \dfrac{2}{5} = \dfrac{42}{25}$.

6

(a) The probability that a train arrives late must be unchanged.

(b) $\dfrac{5!}{5!0!} \times \left(\dfrac{15}{100}\right)^5 \times \left(\dfrac{85}{100}\right)^0 = \text{binompdf}(5, 0.15, 5) = \dfrac{243}{3200000}$.

(c) $P(X=2)+P(X=3)+P(X=4)+P(X=5)=1-(P(X=0)+P(X=1))=$

$$1-\left(\frac{5!}{0!5!}\times\left(\frac{15}{100}\right)^0\times\left(\frac{85}{100}\right)^5+\frac{5!}{1!4!}\times\left(\frac{15}{100}\right)^1\times\left(\frac{85}{100}\right)^4\right)=\frac{16479}{100000}.$$

$\boxed{7}$ $P(X=5)=\dfrac{7!}{5!2!}\times(0.6)^5\times(0.4)^2=\text{binompdf}(7,0.6,5)=0.2612736.$

$\boxed{8}$

(a) 30% must be unchanged in any trial.

(b)

(i) $\dfrac{5!}{3!2!}\times(0.3)^3\times(0.7)^2=0.1323$

(ii)

$P(X=0)+P(X=1)+P(X=2)+P(X=3)$

$=\dfrac{5!}{5!0!}\times(0.3)^0\times(0.7)^5+\dfrac{5!}{4!1!}\times(0.3)^1\times(0.7)^4+\dfrac{5!}{2!3!}\times(0.3)^2\times(0.7)^3+\dfrac{5!}{3!2!}\times(0.3)^3\times(0.7)^2$

$=\text{binomcdf}(5,0.3,3)$

$=\dfrac{48461}{50000}$

$=0.96922$

(c) $5\times\dfrac{30}{100}=\dfrac{15}{10}=1.5$ number of dogs are expected to have skin problems.

(d) $E(X)=1.5$ and $V(X)=5\times0.3\times0.7=\dfrac{21}{20}=1.05$

$\boxed{9}$ Since $n\times\dfrac{1}{4}=200$, we get $n=800$. Hence, $V(X)=800\times\dfrac{1}{4}\times\dfrac{3}{4}=150.$

$\boxed{10}$ Since $np=20$ and $np(1-p)=10$, we get $p=\dfrac{1}{2}$. Hence, $n=40.$

$\boxed{11}$

Part 1. If we normalize the normal graph, $X=120$ corresponds to $z=2$. Using the given information, we notice that the colored area must be 0.977.

Part 2. normalcdf$(-\infty,47,50,4)=0.2266.$

$\boxed{12}$ normalcdf$(13,15,10,4)=0.121.$

$\boxed{13}$ Since $\dfrac{X-12}{4}=1.227$, we get $X\approx16.908.$

$\boxed{14}$ Since $-1.645=\dfrac{X-10}{2}$, we get $X\approx6.71.$

The Essential Guide to Algebra 2

발행 2020년 6월 24일 초판 1쇄

2023년 1월 31일 개정 2쇄

저자 유하림

발행인 최영민

발행처 헤르몬하우스

주소 경기도 파주시 신촌로 16

전화 031 – 8071 – 0088

팩스 031 – 942 – 8688

전자우편 hermonh@naver.com

출판등록 2015년 3월 27일

등록번호 제406 – 2015 – 31호

ISBN 979-11-92520-20-9 (53410)